Solutions and Tests

for

Exploring Creation

with

Biology
2nd Edition

by
Dr. Jay L. Wile and Marilyn F. Durnell

Solutions and Tests for Exploring Creation With Biology, 2nd Edition

Published by
Apologia Educational Ministries, Inc.
1106 Meridian Plaza, Suite 220
Anderson, IN 46016
www.apologia.com

Manufactured in the United States of America
Thirteenth Printing November 2017

ISBN: 978-1-932012-55-2

Printed by LSC Communications

Cover photos: zebra (Dawn Strunc), cell Illustration (© Imagineering / Custom Medical Stock Photo), leaf in background (© Creatas, Inc.), cells in background (© Brand X Pictures)

Exploring Creation With Biology, 2nd Edition
Solutions and Tests

TABLE OF CONTENTS

Teacher's Notes iv

Solutions to the Study Guides:

Solutions To The Study Guide For Module #1 1
Solutions To The Study Guide For Module #2 4
Solutions To The Study Guide For Module #3 7
Solutions To The Study Guide For Module #4 10
Solutions To The Study Guide For Module #5 13
Solutions To The Study Guide For Module #6 16
Solutions To The Study Guide For Module #7 20
Solutions To The Study Guide For Module #8 23
Solutions To The Study Guide For Module #9 30
Solutions To The Study Guide For Module #10 33
Solutions To The Study Guide For Module #11 35
Solutions To The Study Guide For Module #12 38
Solutions To The Study Guide For Module #13 41
Solutions To The Study Guide For Module #14 45
Solutions To The Study Guide For Module #15 48
Solutions To The Study Guide For Module #16 51

Answers for Selected Experiments:

Answers To Experiment 8.1 27
Answers To "Experiment" 8.2 28
Answers To "Experiment" 8.3 29
Answers To Experiment 12.2 40

Answers to the Module Summaries in Appendix B:

Answers To The Summary Of Module #1 55
Answers To The Summary Of Module #2 57
Answers To The Summary Of Module #3 59
Answers To The Summary Of Module #4 61
Answers To The Summary Of Module #5 63
Answers To The Summary Of Module #6 65
Answers To The Summary Of Module #7 66
Answers To The Summary Of Module #8 68
Answers To The Summary Of Module #9 72
Answers To The Summary Of Module #10 74
Answers To The Summary Of Module #11 76
Answers To The Summary Of Module #12 78

Answers To The Summary Of Module #13 80
Answers To The Summary Of Module #14 82
Answers To The Summary Of Module #15 84
Answers To The Summary Of Module #16 86

Module Tests:

Test For Module #1 89
Test For Module #2 91
Test For Module #3 93
Test For Module #4 95
Test For Module #5 97
Test For Module #6 99
Test For Module #7 101
Test For Module #8 103
Test For Module #9 105
Test For Module #10 107
Test For Module #11 109
Test For Module #12 111
Test For Module #13 113
Test For Module #14 115
Test For Module #15 117
Test For Module #16 119

Solutions to the Module Tests:

Solutions To The Test For Module #1 121
Solutions To The Test For Module #2 123
Solutions To The Test For Module #3 125
Solutions To The Test For Module #4 127
Solutions To The Test For Module #5 129
Solutions To The Test For Module #6 131
Solutions To The Test For Module #7 133
Solutions To The Test For Module #8 135
Solutions To The Test For Module #9 138
Solutions To The Test For Module #10 140
Solutions To The Test For Module #11 142
Solutions To The Test For Module #12 144
Solutions To The Test For Module #13 145
Solutions To The Test For Module #14 146
Solutions To The Test For Module #15 148
Solutions To The Test For Module #16 150

Quarterly Tests:

Quarterly Test #1 151
Quarterly Test #2 153
Quarterly Test #3 155
Quarterly Test #4 157

Solutions to the Quarterly Tests:

Solutions To Quarterly Test #1 159
Solutions To Quarterly Test #2 161
Solutions To Quarterly Test #3 164
Solutions To Quarterly Test #4 166

TEACHER'S NOTES
Exploring Creation With Biology, 2nd Edition

Thank you for choosing *Exploring Creation With Biology*. We designed this course to meet the needs of the homeschooling parent. We are very sensitive to the fact that most homeschooling parents do not know biology very well, if at all. As a result, they consider it nearly impossible to teach to their children. This course has several features that make it ideal for such a parent:

1. The course is written in a conversational style. Unlike many authors, we do not get wrapped up in the desire to write formally. As a result, the text is easy to read, and the student feels more like he or she is *learning*, not just reading.

2. The course is completely self-contained. Each module in the student text includes the text of the lesson, experiments to perform, problems to work, and questions to answer. This book contains the solutions to the study guides in the student text, tests, solutions to the tests, and some extra material (answers to several experiments as well as cumulative tests and their solutions).

3. The experiments are designed for the home. Many of them can be done with items that are readily available at either the grocery store or the hardware store. Some of the experiments do require more specialized equipment, however. You can let your budget determine how many of those experiments you will perform. A discussion of the specialized equipment you might consider purchasing for this course is presented in the "Student Notes" section of the student text.

4. Most importantly, this course is Christ-centered. In every way possible, I try to make the science of biology glorify God. One of the most important things that you and your student should get out of this course is a deeper appreciation for the wonder of God's creation!

Pedagogy of the Text

There are two types of exercises that the student is expected to complete: "On Your Own" problems and study guide questions.

- The "On Your Own" problems should be answered as the student reads the text. The act of working out these problems will cement in the student's mind the concepts he or she is trying to learn. The solutions to these problems are included as a part of the student text. The student should feel free to use those solutions to check his work.

- A study guide is found at the end of each module. It is designed to help the student review what has been covered over the course of the module. It should not be started until *after* the student has completed the module. That way, it will function as a review. It can also be used as a study aid for the test. The student should feel free to use the book while answering the study guide questions.

In addition to the problems, there is also a test for each module. Those tests are in this book, but a packet of the tests is also included with this book. You can tear the tests out of the packet and give them to your student so that you need not give him this book to administer the tests. You can also purchase additional packets for additional students. You also have our permission to copy the tests out of this book if you would prefer to do that instead of purchasing additional tests for additional students. **I strongly recommend that you administer each test once the student has completed the module and the study guide. The student should be allowed to have only a calculator, pencil, and paper while taking the test.**

There are also cumulative tests in this book. You can decide whether or not to give these tests to your student. Cumulative tests are probably a good idea if your student is planning to go to college, as he or she will need to get used to taking such tests. There are four cumulative tests along with their solutions. Each cumulative test covers four modules. You have three options as to how you can administer them. You can give each test individually so that the student has four quarterly tests. You can combine the first two quarterly tests and the second two quarterly tests to make two semester tests. You can also combine all four tests to make one end of the year test. If you are giving these tests for the purpose of college preparation, I recommend that you give them as two semester tests, because that is what the student will face in college. The cumulative tests are not in the packet of tests. However, you have our permission to copy them out of this book so that you can give them to your student.

Any information that the student must memorize is centered in the text and put in boldface type. Any boldface words (centered or not) are terms with which the student must be familiar. In addition, all definitions presented in the text need to be memorized. Finally, any information required to answer the questions on the study guide must be committed to memory for the test. If the study guide tells the student that he can refer to a particular table or figure in the text, the test will allow him to do so as well. However, if the study guide does not specifically indicate that the student can reference a figure or table, the student will not be able to reference it for the test.

You will notice that every solution contains an underlined section. That is the answer. The rest is simply an explanation of how to get the answer. For questions that require a sentence or paragraph as an answer, the student need not have *exactly* what is in the solution. The basic message of his or her answer, however, has to be the same as the basic message given in the solution.

Experiments

The experiments in this course are designed to be done as the student is reading the text. I recommend that your student keep a notebook of these experiments. The details of how to perform the experiments and how to keep a laboratory notebook are discussed in the "Student Notes" section of the student text. If you go to the course website that is discussed in the "Student Notes" section of the student text, you will also find examples of how the student should record his or her experiments in the laboratory notebook.

Grading

Grading your student is an important part of this course. I recommend that you *correct* the study guide questions, but I do not recommend that you include the student's score in his or her grade. Instead, I recommend that the student's grade be composed solely of test grades and laboratory notebook grades. Here is what I suggest you do:

1. Give the student a grade for each lab that is done. This grade should not reflect the accuracy of the student's results. Rather, it should reflect how well the student followed directions, how well the student recorded his data, and how well he wrote up the lab in his lab notebook.

2. Give the student a grade for each test. In the test solutions, you will see a point value assigned to each problem. If your student answered the problem correctly, he or she should receive the number of points listed. If your student got a portion of the problem correct, he or she should receive a portion of those points. Your student's percentage grade, then, can be calculated as follows:

$$\text{Student's Grade} = \frac{\text{\# of points received}}{\text{\# of points possible}} \times 100$$

The number of possible points for each test is listed at the bottom of the solutions.

3. The student's overall grade in the course should be weighted as follows: 35% lab grade and 65% test grade. If you use the cumulative tests, make them worth twice as much as each module test. If you really feel that you must include the study guides in the student's total grade, make the labs worth 35%, the tests worth 55%, and the study guides worth 10%. A straight 90/80/70/60 scale should be used to calculate the student's letter grade. This is typical for most schools. If you have your own grading system, please feel free to use it. This grading system is only a suggestion.

Finally, I must tell you that we pride ourselves on the fact that this course is user-friendly and reasonably understandable. At the same time, however, *it is not EASY*. This is a tough course. We have designed it so that any student who gets a "C" or better on the tests will be very well prepared for college.

Question/Answer Service

For all those who use this curriculum, we offer a question/answer service. If there is anything in the modules that you do not understand - from an esoteric concept to a solution for one of the problems - just contact us via any of the methods listed on the **NEED HELP?** page of the student text. You can also contact us regarding any grading issues that you might have. This is our way of helping you and your student to get the maximum benefit from our curriculum.

SOLUTIONS TO THE STUDY GUIDE FOR MODULE #1

1. a. <u>Metabolism</u> – The sum total of all processes in an organism which convert energy and matter from outside sources and use that energy and matter to sustain the organism's life functions

b. <u>Anabolism</u> – The sum total of all processes in an organism which use energy and simple chemical building blocks to produce large chemicals and structures necessary for life

c. <u>Catabolism</u> – The sum total of all processes in an organism which break down chemicals to produce energy and simple chemical building blocks

d. <u>Photosynthesis</u> – The process by which green plants and some other organisms use the energy of sunlight and simple chemicals to produce their own food

e. <u>Herbivores</u> – Organisms that eat only plants

f. <u>Carnivores</u> – Organisms that eat only organisms other than plants

g. <u>Omnivores</u> – Organisms that eat both plants and other organisms

h. <u>Producers</u> – Organisms that produce their own food

i. <u>Consumers</u> – Organisms that eat living producers and/or other consumers for food

j. <u>Decomposers</u> – Organisms that break down the dead remains of other organisms

k. <u>Autotrophs</u> – Organisms that are able to make their own food

l. <u>Heterotrophs</u> – Organisms that depend on other organisms for their food

m. <u>Receptors</u> – Special structures that allow living organisms to sense the conditions of their internal or external environment

n. <u>Asexual reproduction</u> – Reproduction accomplished by a single organism

o. <u>Sexual reproduction</u> – Reproduction that requires two organisms

p. <u>Inheritance</u> – The process by which physical and biological characteristics are transmitted from the parent (or parents) to the offspring

q. <u>Mutation</u> – An abrupt and marked change in the DNA of an organism compared to that of its parents

r. <u>Hypothesis</u> – An educated guess that attempts to explain an observation or answer a question

s. <u>Theory</u> – A hypothesis that has been tested with a significant amount of data

t. <u>Scientific law</u> – A theory that has been tested by and is consistent with generations of data

u. <u>Microorganisms</u> – Living creatures that are too small to see with the naked eye

v. <u>Abiogenesis</u> – The idea that long ago, very simple life forms spontaneously appeared through chemical reactions

w. <u>Prokaryotic cell</u> – A cell that has no distinct, membrane-bounded organelles

x. <u>Eukaryotic cell</u> – A cell with distinct, membrane-bounded organelles

y. <u>Species</u> – A unit of one or more populations of individuals that can reproduce under normal conditions, produce fertile offspring, and are reproductively isolated from other such units

z. <u>Taxonomy</u> – The science of classifying organisms

aa. <u>Binomial nomenclature</u> – Naming an organism with its genus and species name

2. The four criteria for life are:

1. <u>All life forms contain deoxyribonucleic acid, which is called DNA.</u>

2. <u>All life forms have a method by which they extract energy from the surroundings and convert it into energy that sustains them.</u>

3. <u>All life forms can sense changes in their surroundings and respond to those changes.</u>

4. <u>All life forms reproduce.</u>

3. Carnivores eat non-plants. This means they depend on other organisms for food, making them <u>heterotrophs</u>, which are also known as <u>consumers</u>.

4.   If the tentacles are cut off, then the organism has no receptors, which sense the conditions of the environment. Thus, <u>sensing changes in the surroundings and responding to those changes will be hard for this wounded creature.</u>

5. <u>These organisms reproduce sexually.</u> In sexual reproduction, the offspring's traits are a blend of the parents, their parents, and so on. This would account for the differences between parent and offspring.

6. <u>Science cannot prove anything.</u> Since it is based on experiments that may be flawed, its conclusions are always tentative.

7. <u>In the scientific method, a person starts by making observations. The person then develops a hypothesis to explain those observations or to answer a question. The person (often with the help of others) then designs experiments to test the hypothesis. After the hypothesis has been tested by a significant amount of data and is consistent with all of it, then it becomes theory. After more testing with generations of data, the theory could become a scientific law.</u>

8. <u>The story of spontaneous generation shows how almost 2,000 years of executing the scientific method resulted in a law that was clearly wrong.</u> Thus, you can't put too much faith in scientific laws. They are fallible.

9. <u>The wise person trusts the Bible</u>, because it is infallible.

10. <u>Abiogenesis is a theory that states that life sprang from non-living chemicals eons ago. This is an example of spontaneous generation, a former law that said life could arise from non-life.</u> We now know that this law is wrong.

11. <u>Kingdom, Phylum, Class, Order, Family, Genus, Species</u>

12. <u>Animalia</u> - Since it is multicellular, it is not Monera or Protista. In addition, it is not Plantae because it is not an autotroph (consumers are heterotrophs), and it is not Fungi because it is not a decomposer.

13. Since it has eukaryotic cells, it would be in the <u>Eukarya</u> domain.

14. It belongs in kingdom <u>Monera</u>, because all organisms made of prokaryotic cells belong to this kingdom.

15. All members of kingdom Monera are either in the Archaea domain or the Bacteria domain. <u>You cannot tell which domain without knowing more about the organism. However, it is either in Archaea or Bacteria, depending on its characteristics.</u>

16. a.   1. macroscopic, proceed to key 3
       3. heterotrophic, proceed to key 5
       5. consumer, which means <u>kingdom Animalia</u>, proceed to key 6.
       6. backbone, which means <u>phylum Chordata</u>, proceed to key 22
     22. beak, proceed to key 23
     23. no scales, proceed to key 26
     26. feathers, proceed to key 28
     28. feathers, which means <u>class Aves</u>

  b.   1. macroscopic, proceed to key 3
       3. heterotrophic, proceed to key 5
       5. consumer, which means <u>kingdom Animalia</u>, proceed to key 6.
       6. no backbone, proceed to key 7
       7. right and left sides, proceed to key 9
       9. external plates, which means <u>phylum Arthropoda</u>, proceed to key 14
     14. three pairs of walking legs, which means <u>class Insecta</u>, proceed to key 16
     16. wings, proceed to key 17
     17. all wings transparent, proceed to key 18
     18. cannot sting, which means <u>order Diptera</u>

SOLUTIONS TO THE STUDY GUIDE FOR MODULE #2

1. a. <u>Pathogen</u> – An organism that causes disease

b. <u>Saprophyte</u> – An organism that feeds on dead matter

c. <u>Parasite</u> – An organism that feeds on a living host

d. <u>Aerobic organism</u> – An organism that requires oxygen

e. <u>Anaerobic organism</u> – An organism that does not require oxygen

f. <u>Steady state</u> – A state in which members of a population die as quickly as new members are born

g. <u>Exponential growth</u> – Population growth that is unhindered because of the abundance of resources for an ever-increasing population

h. <u>Logistic growth</u> – Population growth that is controlled by limited resources

i. <u>Conjugation</u> – A temporary union of two organisms for the purpose of DNA transfer

j. <u>Plasmid</u> – A small, circular section of extra DNA that confers one or more traits to a bacterium and can be reproduced separately from the main bacterial genetic code

k. <u>Transformation</u> – The transfer of a DNA segment from a nonfunctional donor cell to that of a functional recipient cell

l. <u>Transduction</u> – The process in which infection by a virus results in DNA being transferred from one bacterium to another

m. <u>Endospore</u> – The DNA and other essential parts of a bacterium coated with several hard layers

n. <u>Strains</u> – Organisms from the same species that have markedly different traits

2. a. <u>plasma membrane</u> b. <u>flagellum</u> c. <u>capsule</u> d. <u>DNA</u> e. <u>cytoplasm</u> f. <u>cell wall</u> g. <u>fimbria</u> h. <u>ribosome</u>

3. a. Plasma membrane: <u>To negotiate what materials pass into and out of the cell</u>

b. Flagellum: <u>To move the bacterium from place to place</u>

c. Capsule: <u>To adhere to surfaces as well as to ward off infection-fighting agents</u>

d. DNA: <u>To store the information needed to make an organism a living thing</u>

e. Cytoplasm: <u>To hold the DNA and ribosomes in place</u>

f. Cell wall: <u>To keep the interior of the cell together and to hold the cell's shape</u>

g. Fimbria: <u>To grasp onto surfaces or another bacterium during conjugation</u>

h. Ribosome: <u>To make proteins</u>

4. Most bacteria are <u>heterotrophic decomposers</u>.

5. Parasites, by definition, feed on something produced by a host. They therefore cannot make their own food. This makes them <u>heterotrophic</u>.

6. At first, <u>the DNA loop attaches to a point on the plasma membrane</u>. After that, <u>the DNA is copied, and the copy is attached to a point on the plasma membrane near the original</u>. Then, <u>the cell wall elongates</u>, which separates the two loops of DNA. Once they are sufficiently separated, <u>new cell wall and plasma membrane material grow, closing the two loops off from each other. Eventually, the cell wall and plasma membrane pinch down</u>, forming two cells where there was only one before.

7. <u>The bacteria that are there were in the food as endospores.</u> The bacteria that were in the food before it was dehydrated formed endospores to survive a little while without water. When water was added to the food, the conditions were once again favorable for bacteria, so the cells burst from the endospores.

8. If the environment is rich with resources and there are only a few bacteria, the population will experience <u>exponential growth</u>, because there is nothing to limit their reproduction.

9. Since the growth is logistic, <u>the resources are limited but do no run out</u>. If the resources ran out, the population would experience a decline, and that is not a part of logistic growth.

10. <u>Genetic recombination can pass a trait from one bacterium to another. If that trait allows the recipient to survive conditions that it otherwise wouldn't, the population is affected, because the recipient continues to live and reproduce asexually.</u>

11. <u>Coccus</u> - Spherical
   <u>Bacillus</u> - Rod-shaped
   <u>Spirillum</u> - Helical

12. If it is Gram-negative, it is in <u>phylum Gracilicutes</u>. In this phylum, two classes contain photosynthetic bacteria, which are autotrophs. Since this bacterium is heterotrophic, it must belong to the only other class, <u>class Scotobacteria.</u>

13. Gram-positive means <u>phylum Firmicutes</u>. Since it is spirillum-shaped, it is neither coccus nor bacillus. Thus, it is in <u>class Thallobacteria</u>.

14. Bacteria with no cell walls belong to the <u>phylum Tenericutes</u>, which has only one class, <u>class Mollicutes</u>.

15. This bacterium most likely is in phylum <u>Mendosicutes</u> and <u>class Archaebacteria</u>, because bacteria with exotic cell walls can survive conditions that other organisms cannot.

16. In the six-kingdom system, kingdom Monera is split into <u>kingdom Archaebacteria</u> and <u>kingdom Eubacteria</u>.

17. To grow and reproduce, ideal conditions for most bacteria include: <u>Moisture, moderate temperatures, nutrition, darkness, and the proper amount of oxygen</u>.

18. To reduce the chance of bacterial infection you can:

a) <u>Heat the food so that most bacteria die and then seal it away from fresh air</u>.
b) <u>Dehydrate the food</u>.
c) <u>Freeze the food</u>.
d) <u>Pasteurize the food.</u>
e) <u>Keep it in the refrigerator</u>.

SOLUTIONS TO THE STUDY GUIDE FOR MODULE #3

1. a. <u>Pseudopod</u> - A temporary, foot-like extension of a cell, used for locomotion or engulfing food

b. <u>Nucleus</u> – The region of a eukaryotic cell that contains the cell's main DNA

c. <u>Vacuole</u> – A membrane-bounded "sac" within a cell

d. <u>Ectoplasm</u> – The thin, watery cytoplasm near the plasma membrane of some cells

e. <u>Endoplasm</u> – The dense cytoplasm found in the interior of many cells

f. <u>Flagellate</u> – A protozoan that propels itself with a flagellum

g. <u>Pellicle</u> – A firm, flexible coating outside the plasma membrane

h. <u>Chloroplast</u> – An organelle containing chlorophyll for photosynthesis

i. <u>Chlorophyll</u> – A pigment necessary for photosynthesis

j. <u>Eyespot</u> – A light-sensitive region in certain protozoa

k. <u>Symbiosis</u> – A close relationship between two or more species where at least one benefits

l. <u>Mutualism</u> – A relationship between two or more organisms of different species where all benefit from the association

m. <u>Commensalism</u> – A relationship between two organisms of different species where one benefits and the other is neither harmed nor benefited

n. <u>Parasitism</u> – A relationship between two organisms of different species where one benefits and the other is harmed

o. <u>Cilia</u> – Hairlike projections that extend from the plasma membrane and are used for locomotion

p. <u>Spore</u> – A reproductive cell with a hard, protective coating

q. <u>Plankton</u> – Tiny organisms that float in the water

r. <u>Zooplankton</u> – Tiny floating organisms that are either small animals or protozoa

s. <u>Phytoplankton</u> – Tiny floating photosynthetic organisms, primarily algae

t. <u>Thallus</u> – The body of a plant-like organism that is not divided into leaves, roots, or stems

u. <u>Cellulose</u> – A substance (made of sugars) that is common in the cell walls of many organisms

v. <u>Holdfast</u> – A special structure used by an organism to anchor itself

w. Sessile Colony – A colony that uses holdfasts to anchor itself to an object

2. There is no real answer for this question. Just be sure that you can name the subkingdom and phylum of each organism in Figure 3.1 when you see its picture.

3. *Euglena* and *Spirogyra*. Each of these organisms use chlorophyll for photosynthesis and thus have chloroplasts. The other two genera contain exclusively heterotrophic organisms, which obviously do not use photosynthesis.

4. A contractile vacuole collects excess water in a cell and releases it into the surroundings to reduce the pressure inside the cell. This keeps the cell from exploding. The food vacuole, on the other hand, stores food while it is being digested and has nothing to do with excess water or pressure.

5. Endoplasm is thick, while ectoplasm is thin and watery. Endoplasm is found in the central region of the cell, while ectoplasm is found near the plasma membrane.

6. The amoeba uses pseudopods which it creates by deforming its body. The euglena, on the other hand uses a flagellum. There is one bit of similarity. When it wants to move quickly, the euglena deforms its body in an almost earthworm-type motion. This is used to enhance the motion supplied by the flagellum, and is something like the amoeba's motion.

7. There are more than three, but you only need to recall three. The ones we discussed in this module are *Entamoeba histolytica, Trypanosoma, Balantidium coli, Plasmodium,* and *Toxoplasma*.

8. Sarcodina: pseudopods, Mastigophora: flagella, Ciliophora: cilia.

9. These organisms form spores as a natural part of their life cycle and have no real means of locomotion.

10. *Trichonympha* is an example of mutualism, because both the *Trichonympha* and the termite benefit from the situation. The tapeworm is an example of parasitism, since only the tapeworm benefits. The host is hurt by the situation.

11. Ciliates require so much energy that they must have a nucleus (called the macronucleus) devoted solely to metabolism. The other, smaller nucleus (the micronucleus) controls reproduction.

12. In conjugation between paramecia, there is a mutual exchange of DNA so that each paramecium gets new DNA. We learned in Module #2 that when bacteria conjugate, only one bacterium (the recipient) gets new DNA.

13. Spores are formed as a natural part of an organism's lifestyle. Cysts, however, are only formed in the case of life-threatening conditions. If those conditions do not exist, cysts will not be formed. Thus, the first group produced cysts. The second group produced spores, making them a part of phylum Sporozoa.

14. A euglena can either live on the dead remains of other organisms or it can produce its own food by photosynthesis. This combination of autotrophic and heterotrophic behavior is rather unique in God's creation.

15. <u>Phylum Chrysophyta contains the diatoms</u>, which are responsible for most of the world's photosynthesis.

16. In the answers below, we list all of the phyla that apply. You only need to list one.

<u>Food vacuole - purpose: store food, phyla: Sarcodina, Mastigophora, Ciliophora</u>

<u>Contractile vacuole - purpose: remove excess water, reducing pressure, phyla: Sarcodina, Mastigophora, Ciliophora</u>

<u>Flagellum - purpose: locomotion, phylum: Mastigophora, Pyrrophyta</u>

<u>Pellicle - purpose: retains cell shape, phyla: Mastigophora, Ciliophora</u>

<u>Chloroplast - purpose: stores chlorophyll, phylum: Chlorophyta or Mastigophora</u>

<u>Eyespot - purpose: detects light, phylum: Mastigophora</u>

<u>Cilia - purpose: locomotion, phylum: Ciliophora</u>

<u>Nucleus - purpose: contains DNA, phyla: all phyla in Protista</u>

<u>Oral groove - purpose: food intake and conjugation, phylum: Ciliophora</u>

17. <u>These deposits are called diatomaceous earth and are used as abrasives and filters.</u>

18. <u>A red tide is an algae bloom of dinoflagellates</u>, which belong to phylum Pyrrophyta.

19. <u>Phaeophyta and Rhodophyta</u>

20. Members of phylum Phaeophyta have <u>alginic acid (or just algin)</u> in their cell walls. This is the thickening agent used in the foods listed.

SOLUTIONS TO THE STUDY GUIDE FOR MODULE #4

1. a. <u>Extracellular digestion</u> – Digestion that takes place outside of the cell

b. <u>Mycelium</u> – The part of the fungus responsible for extracellular digestion and absorption of the digested food

c. <u>Hypha</u> – A filament of fungal cells

d. <u>Rhizoid hypha</u> – A hypha that is imbedded in the material on which the fungus grows

e. <u>Aerial hypha</u> – A hypha that is not imbedded in the material upon which the fungus grows

f. <u>Sporophore</u> – Specialized aerial hypha that produces spores

g. <u>Stolon</u> – An aerial hypha that asexually reproduces to make more filaments

h. <u>Haustorium</u> – A hypha of a parasitic fungus that enters the host's cells, absorbing nutrition directly from the cytoplasm

i. <u>Chitin</u> – A chemical that provides both toughness and flexibility

j. <u>Membrane</u> – A thin covering of tissue

k. <u>Fermentation</u> – The anaerobic breakdown of sugars into smaller molecules

l. <u>Zygospore</u> – A zygote surrounded by a hard, protective covering

m. <u>Zygote</u> – The result of sexual reproduction when each parent contributes half of the DNA necessary for the offspring

n. <u>Antibiotic</u> – A chemical secreted by a living organism that kills or reduces the reproduction rate of other organisms

2. Characteristics common to the majority of fungi were discussed in the section entitled "General Characteristics of Fungi." It was noted, however, that of the specialized hyphae, only rhizoid hyphae are common to the vast majority of fungi.

<u>Common to the majority of fungi</u>	<u>Present in only some</u>
extracellular digestion	stolons (specialized hyphae)
chitin	caps and stalks (only mushrooms have them)
mycelia	sporangiophores (specialized hyphae with enclosed spores)
hyphae	haustoria (specialized hyphae)
cells (all living creatures have them)	motile spores (Chytridiomycota and some slime molds)
rhizoid hyphae	septate hyphae (many have non-septate hyphae)

3. <u>Typically, we see only the fruiting body of a mushroom. Like an iceberg, that visible part is only a small fraction of the total mushroom, because the mycelium is the largest component of a mushroom.</u>

4. <u>Septate hyphae have cell walls to separate the cells while non-septate hyphae do not.</u>

5. <u>Rhizoid hyphae support the fungus and digest the food; a stolon asexually reproduces; a sporophore releases spores for reproduction; a haustorium invades the cells of a living host to absorb food directly from the cytoplasm.</u>

6. <u>Stolons and sporophores are aerial.</u> Aerial hyphae are not imbedded in the material upon which the fungus grows. In order to perform their jobs, rhizoid hyphae and haustoria must be imbedded in the material.

7. <u>A sporangiophore produces its spores in an enclosure; a conidiophore does not.</u>

8. Basidiomycota: <u>Form sexual spores on club-like basidia</u>
   Ascomycota: <u>Form sexual spores in sac-like asci</u>
   Zygomycota: <u>Form sexual spores where hyphae fuse</u>
   Chytridiomycota: <u>Form spores with flagella</u>
   Deuteromycota: <u>Fungi with no known method of sexual reproduction</u>
   Myxomycota: <u>Fungi that look like protozoa for much of their lives</u>

9. <u>A mushroom begins life as a small mycelium that grows from spores which have come from another mushroom. As the mycelium begins to grow, it might encounter a compatible mycelium. As the two mycelia begin to intertwine, their hyphae will sexually reproduce. Eventually, the newly-produced hyphae will form a complex web and enclose themselves in a membrane. When the hyphae are formed in the membrane, we say that the mushroom has reached the button stage of its existence. At that point, the hyphae begin filling with water quickly, and eventually the stipe and cap (the fruiting body) of the mushroom break through the membrane. The fruiting body of the mushroom releases its spores, which will grow into new mycelia if they land in suitable habitats.</u>

10. The main difference is where they form their spores. <u>Mushrooms form spores on basidia that exist in the gills of the cap, puffballs produce spores on basidia enclosed in a membrane, and shelf fungi produce spores on basidia in pores on the fruiting body.</u>

11. <u>An alternate host is used by a parasitic fungus at some stage in its life. It is not the host that the fungus spends most of its life on</u>; it is simply a temporary host that is necessary for a certain part of the fungus' development. <u>Rusts use alternate hosts.</u>

12. <u>Yeast are best known for fermentation. They belong to phylum Ascomycota.</u>

13. <u>In budding, the offspring stays attached to the parent until it has grown. In bacterial asexual reproduction, the offspring grows on its own.</u>

14. There are many pathogenic fungi. You need only list two:

1) rusts - crop damage   4) *Cryphonectria parasitica* - chestnut blight
2) smuts - crop damage   5) *Ophiostoma ulmi* - Dutch elm disease
3) ergot of rye (*Claviceps purpurea*) - death 6) *Synchytrium endobioticum* – potato wart

15. Bread mold can asexually reproduce when a stolon elongates and eventually starts another mycelium. It can also asexually reproduce when an aerial hypha forms a sporophore (typically a sporangiophore). Sexually, bread molds reproduce when two mycelia form a zygospore.

16. If we do not know what its sexual mode of spore formation is, we place the fungus in phylum Deuteromycota.

17. If an antibiotic is used too much, resistant strains of the pathogen it is supposed to destroy can be formed.

18. Penicillin is extracted from a fungus in genus *Penicillium*.

19. In its feeding stage, a slime mold is a plasmodium. During that time, it resembles organisms from kingdom Protista.

20. Slime molds must have water to survive. Keep the habitat dry, and all slime molds will die.

21. Fungi participate in mutualism by forming lichens and mycorrhizae. A lichen is a mutualistic relationship between a fungus and an alga. The alga produces food for both creatures via photosynthesis, and the fungus supports and protects the alga. Mycorrhizae are mutualistic relationships between a fungus' mycelium and a plant's root system. The mycelium takes nutrients from the root while it collects minerals from the soil and gives them to the root.

22. A soredium is the specialized spore produced by most lichens. It contains spores for both the fungus and the alga.

SOLUTIONS TO THE STUDY GUIDE FOR MODULE #5

1. a. <u>Matter</u> – Anything that has mass and takes up space

b. <u>Model</u> – An explanation or representation of something that cannot be seen

c. <u>Element</u> – A collection of atoms that all have the same number of protons

d. <u>Molecules</u> – Chemicals that result from atoms linking together

e. <u>Physical change</u> – A change that affects the appearance but not the chemical makeup of a substance

f. <u>Chemical change</u> – A change that alters the makeup of the elements or molecules of a substance

g. <u>Phase</u> – One of three forms - solid, liquid, or gas - which every substance is capable of attaining

h. <u>Diffusion</u> – The random motion of molecules from an area of high concentration to an area of low concentration

i. <u>Concentration</u> – A measurement of how much solute exists within a certain volume of solvent

j. <u>Semipermeable membrane</u> – A membrane that allows some molecules to pass through but does not allow other molecules to pass through

k. <u>Osmosis</u> – The tendency of a solvent to travel across a semipermeable membrane into areas of higher solute concentration

l. <u>Catalyst</u> – A substance that alters the speed of a chemical reaction but is not used up in the process

m. <u>Organic molecule</u> – A molecule that contains only carbon and any of the following: hydrogen, oxygen, nitrogen, sulfur, and/or phosphorous

n. <u>Biosynthesis</u> – The process by which living organisms produce larger molecules from smaller ones

o. <u>Isomers</u> – Two different molecules that have the same chemical formula

p. <u>Monosaccharides</u> – Simple carbohydrates that contain 3 to 10 carbon atoms

q. <u>Disaccharides</u> – Carbohydrates that are made up of two monosaccharides

r. <u>Polysaccharides</u> – Carbohydrates that are made up of more than two monosaccharides

s. <u>Dehydration reaction</u> – A chemical reaction in which molecules combine by removing water

t. <u>Hydrolysis</u> – Breaking down complex molecules by the chemical addition of water

u. <u>Hydrophobic</u> – Lacking any affinity to water

v. <u>Saturated fat</u> – A lipid made from fatty acids that have no double bonds between carbon atoms

w. <u>Unsaturated fat</u> – A lipid made from fatty acids that have at least one double bond between carbon atoms

x. <u>Peptide bond</u> – A bond that links amino acids together in a protein

y. <u>Hydrogen bond</u> – A strong attraction between hydrogen atoms and certain other atoms (usually oxygen or nitrogen) in specific molecules

2. <u>In an atom, protons and neutrons cluster together at the center, which is called the nucleus. Electrons orbit around the nucleus.</u>

3. <u>The number of electrons (or protons) in an atom determines the vast majority of its characteristics.</u>

4. <u>When a number appears after an atom's name, it tells you the sum of protons and neutrons in the atom's nucleus.</u>

5. <u>An element contains all atoms that have the same number of protons (and therefore the same number of electrons), regardless of the number of neutrons. An atom is a single entity, determined by its number of protons, electrons, *and* neutrons.</u>

6. Since atoms have the same number of electrons and protons, there must be <u>32 electrons</u>.

7. The subscripts after the elemental abbreviations tell you how many of each atom is in the molecule. Thus, there are <u>3 carbons, 8 hydrogens, and 1 oxygen, for a grand total of 12 atoms.</u>

8. a. <u>Molecule</u>, because it has several atoms linked together
b. <u>Atom</u>, because it specifies number of neutrons and protons
c. <u>Element</u>, because it is by itself but does not specify the number of neutrons and protons

9. Adding energy causes molecules to go from solid to liquid to gas. Thus, <u>the liquid will turn into a gas</u>. To turn it into a solid, you must take energy from it.

10. <u>A semipermeable membrane should *not* be used</u>. For diffusion to work, both solute and solvent must be able to travel across the membrane. Semipermeable membranes typically allow only solvent molecules to pass.

11. Since the water levels changed, that means solvent traveled from one side of the membrane to the other, but solute did not. <u>This is osmosis, which requires a semipermeable membrane</u>.

12. a. Reactants appear to the left of the arrow. The number to the left of the chemical formulas, however, do not describe the reactants. Instead, they tell you how many of each reactant molecule. Thus, the reactants are <u>N_2 and H_2</u>.

b. Products appear on the right side of the arrow. The product is <u>NH_3</u>.

c. There are <u>three</u> H_2 molecules in the reaction, because of the "3" to the left of H_2.

13. Photosynthesis is represented by:

$$6CO_2 \ + \ 6H_2O \ \rightarrow \ C_6H_{12}O_6 + 6O_2$$

In order for a plant to carry out photosynthesis, it needs <u>CO_2, H_2O, energy from sunlight, and a catalyst like chlorophyll</u>.

14. <u>Reactions can also be sped up by increasing temperature.</u>

15. Carbohydrates have carbon, hydrogen, and oxygen and no other elements. In addition, like water, they must have twice as many H's as O's. Only molecule <u>d</u> fits that bill.

16. <u>Dehydration reactions build up these molecules, and hydrolysis reactions can break them down,</u> providing the proper enzyme exists.

17. An acid must contain an acid group, which looks like this:

$$\begin{array}{c} O \\ \| \\ -C-OH \end{array}$$

Only molecule <u>c</u> has one.

18. <u>The pH scale measures the acidity or alkalinity of a solution. On this scale, 7 is neutral. Lower than 7 pH's are acidic, and higher than 7 are alkaline. The lower the pH the more acidic, and the higher the pH the more alkaline.</u>

19. <u>Amino acids link together to make proteins, fatty acids link to glycerol to make lipids, and monosaccharides link together to make polysaccharides.</u>

20. <u>These two proteins will not have the same properties.</u> Not only the number and type but also the order of amino acids determine a protein's structure and function.

21. <u>Enzymes are a special class of proteins that are used as catalysts.</u>

22. <u>The "lock and key" theory of enzyme action says that an enzyme has an active site that is shaped especially for the molecule that it must work on. The action that the enzyme takes cannot happen until the molecule attaches to that active site.</u> Since the active site is shaped for a specific molecule, such an enzyme cannot work on other molecules, unless it happens to have the same shape.

23. <u>The three basic parts of a nucleotide are the phosphate group, the sugar, and the base.</u>

24. <u>DNA stores information as a sequence of nucleotide bases,</u> much like all of the English language can be stored as a sequence of dots and dashes in Morse code.

25. <u>Hydrogen bonds between the nucleotide bases hold the two helixes of DNA together.</u>

SOLUTIONS TO THE STUDY GUIDE FOR MODULE #6

1. a. <u>Absorption</u> – The transport of dissolved substances into cells

b. <u>Digestion</u> – The breakdown of absorbed substances

c. <u>Respiration</u> – The breakdown of food molecules with a release of energy

d. <u>Excretion</u> – The removal of soluble waste materials

e. <u>Egestion</u> – The removal of nonsoluble waste materials

f. <u>Secretion</u> – The release of biosynthesized substances

g. <u>Homeostasis</u> – Maintaining the status quo

h. <u>Reproduction</u> – Producing more cells

i. <u>Cytology</u> – The study of cells

j. <u>Cell wall</u> – A rigid structure on the outside of certain cells, usually plant and bacteria cells

k. <u>Middle lamella</u> – The thin film between the cell walls of adjacent plant cells

l. <u>Plasma membrane</u> – The semipermeable membrane between the cell contents and either the cell wall or the cell's surroundings

m. <u>Cytoplasm</u> – A jellylike fluid inside the cell in which the organelles are suspended

n. <u>Ions</u> – Substances in which at least one atom has an imbalance of protons and electrons

o. <u>Cytoplasmic streaming</u> – The motion of cytoplasm in a cell that results in a coordinated movement of the cell's contents

p. <u>Mitochondria</u> – The organelles in which nutrients are converted to energy

q. <u>Lysosome</u> – The organelle in animal cells responsible for hydrolysis reactions that break down proteins, polysaccharides, disaccharides, and some lipids

r. <u>Ribosomes</u> – Non-membrane-bound organelles responsible for protein synthesis

s. <u>Endoplasmic reticulum</u> – An organelle composed of an extensive network of folded membranes that performs several tasks within a cell

t. <u>Rough ER</u> – ER that is dotted with ribosomes

u. <u>Smooth ER</u> – ER that has no ribosomes

v. <u>Golgi bodies</u> – The organelles in which proteins and lipids are stored and then modified to suit the needs of the cell

w. <u>Leucoplasts</u> – Organelles that store starches or oils

x. <u>Chromoplasts</u> – Organelles that contain pigments used in photosynthesis

y. <u>Central vacuole</u> – A large vacuole that rests at the center of most plant cells and is filled with a solution that contains a high concentration of solutes

z. <u>Waste vacuoles</u> – Vacuoles that contain the waste products of digestion

aa. <u>Phagocytosis</u> – The process by which a cell engulfs foreign substances or other cells

bb. <u>Phagocytic vacuole</u> – A vacuole that holds the matter which a cell engulfs

cc. <u>Pinocytic vesicle</u> – Vesicle formed at the plasma membrane to allow the absorption of large molecules

dd. <u>Secretion vesicle</u> – Vesicle that holds secretion products so that they can be transported to the plasma membrane and released

ee. <u>Microtubules</u> – Spiral strands of protein molecules that form a tubelike structure

ff. <u>Nuclear membrane</u> – A highly-porous membrane that separates the nucleus from the cytoplasm

gg. <u>Chromatin</u> – Clusters of DNA, RNA, and proteins in the nucleus of a cell

hh. <u>Cytoskeleton</u> – A network of fibers that holds the cell together, helps the cell to keep its shape, and aids in movement

ii. <u>Microfilaments</u> – Fine, threadlike proteins found in the cell's cytoskeleton

jj. <u>Intermediate filaments</u> – Threadlike proteins in the cell's cytoskeleton that are roughly twice as thick as microfilaments

kk. <u>Phospholipid</u> – A lipid in which one of the fatty acid molecules has been replaced by a molecule that contains a phosphate group

ll. <u>Passive transport</u> – Movement of molecules through the plasma membrane according to the dictates of osmosis or diffusion

mm. <u>Active transport</u> – Movement of molecules through the plasma membrane (typically opposite the dictates of osmosis or diffusion) aided by a process that requires energy

nn. <u>Isotonic solution</u> – A solution in which the concentration of solutes is essentially equal to that of the cell that resides in the solution

oo. Hypertonic solution – A solution in which the concentration of solutes is greater than that of the cell that resides in the solution

pp. Plasmolysis – Collapse of a walled cell's cytoplasm due to a lack of water

qq. Cytolysis – The rupturing of a cell due to excess internal pressure

rr. Hypotonic solution – A solution in which the concentration of solutes is less than that of the cell which resides in the solution

ss. Activation energy – Energy necessary to get a chemical reaction going

2. The ribosome makes proteins, the smooth ER and rough ER both make molecules like polysaccharides and lipids, the Golgi bodies package products of biosynthesis, the chloroplasts are involved in photosynthesis, the leucoplasts store products of biosynthesis, and the nucleus participates in the production of proteins through the DNA that it contains. You could also add secretion vesicles, as they move products of biosynthesis to the plasma membrane for secretion. You could add cytoplasm as well, but that is not generally considered an organelle.

3. The cytoskeleton and the endoplasmic reticulum help the cell hold its shape.

4. The cell has a central vacuole that expands as the cell absorbs water. This causes turgor pressure in the cell, which counteracts osmosis.

5. It is an animal cell. Plant cells do not have those organelles. Some plant cells have centrioles, but not very many.

6. The secretion product must go to the Golgi bodies to be packaged, at which point it is put in a secretion vesicle. It must then travel to the plasma membrane, where it is released. If the cell has a cell wall, it must pass through the cell wall as well. Once again, you could put cytoplasm in here as well.

7. The cytoplasm does cytoplasmic streaming, the smooth ER and rough ER deal with movement, the Golgi bodies package things for movement, secretion vesicles and waste vacuoles move things to the plasma membrane, the centrioles produce microtubules which produce movement, and the cytoskeleton as a whole is involved with movement. You could add the plasma membrane here, as it deals with the movement of things into and out of the cell.

8. As discussed in the first section, the are: absorption, digestion, respiration, biosynthesis, excretion, egestion, secretion, movement, irritability, homeostasis, and reproduction.

9. The plasma membrane is composed of phospholipids, cholesterol, and proteins. There are carbohydrates attached to certain proteins (making them glycoproteins) and lipids (making them glycolipids), but they are considered a part of the glycoprotein or glycolipid to which they are attached.

10. A phospholipid has two fatty acid molecules and a small molecule with a phosphate group, whereas a normal lipid just has 3 fatty acid molecules. This makes the phospholipid have a hydrophilic end, which the regular lipid does not.

11. <u>Since the phospholipids have a hydrophilic end and a hydrophobic end, they always "know" how to reassemble</u>.

12. Active transport requires energy from the cell, whereas passive transport does not. Thus, the <u>active transport</u> would slow down.

13. a. <u>phospholipid</u> (You could be very precise and say the hydrophilic end of a phospholipid.)
b. <u>protein</u> (You could say active transport site) c. <u>glycoprotein</u> d. <u>carbohydrate</u> e. <u>cholesterol</u>
f. <u>filaments of the cytoskeleton</u> g. <u>glycolipid</u>

14. Since it died by implosion, the cell lost water. Water is lost by osmosis when the cell is in a solution which has a higher concentration of solutes than the inside of the cell. Thus, this was a <u>hypertonic</u> solution.

15. The four stages are: <u>glycolysis (two ATPs), the formation of acetyl coenzyme A (no ATPs), the Krebs cycle (two ATPs), and the electron transport system (32 ATPs)</u>.

16. <u>ATP supplies a package for the energy produced in cellular respiration. It releases its energy gently, so that the energy does not destroy the cell</u>.

17. The only stage that can run is <u>glycolysis</u>. After that, the cell forms lactic acid or alcohol with the products. The cell can only make <u>two ATPs</u> per molecule of glucose this way.

18. <u>With no ADP, the cell will not be able to make ATP in which to store the energy from cellular respiration. Thus, the cell could make energy, but it could never use the energy</u>!

19. <u>The lysosome</u> performs hydrolysis which breaks down large molecules (like polysaccharides) into small molecules (like monosaccharides).

SOLUTIONS TO THE STUDY GUIDE FOR MODULE #7

1. a. <u>Genetics</u> – The science that studies how characteristics get passed from parent to offspring

b. <u>Genetic factors</u> – The general guideline of traits determined by a person's DNA

c. <u>Environmental factors</u> – Those "nonbiological" factors that are involved in a person's surroundings such as the nature of the person's parents, the person's friends, and the person's behavioral choices

d. <u>Spiritual factors</u> – The factors in a person's life that are determined by the quality of his or her relationship with God

e. <u>Gene</u> – A section of DNA that codes for the production of a protein or a portion of protein, thereby causing a trait

f. <u>Messenger RNA</u> – The RNA that performs transcription

g. <u>Anticodon</u> – A three-nucleotide base sequence on tRNA

h. <u>Codon</u> – A sequence of three nucleotide bases on mRNA that refers to a specific type of amino acid

i. <u>Chromosome</u> – DNA coiled around and supported by proteins, found in the nucleus of the cell

j. <u>Mitosis</u> – A process of asexual reproduction in eukaryotic cells

k. <u>Interphase</u> – The time interval between cellular reproduction

l. <u>Centromere</u> – The region that joins two sister chromatids

m. <u>Mother cell</u> – A cell ready to begin reproduction, containing duplicated DNA and centrioles

n. <u>Karyotype</u> – The figure produced when the chromosomes of a species during metaphase are arranged according to their homologous pairs

o. <u>Diploid cell</u> – A cell with chromosomes that come in homologous pairs

p. <u>Haploid cell</u> – A cell that has only one representative of each chromosome pair

q. <u>Diploid number (2n)</u> – The total number of chromosomes in a diploid cell

r. <u>Haploid number (n)</u> – The number of homologous pairs in a diploid cell

s. <u>Meiosis</u> – The process by which a diploid (2n) cell forms gametes (n)

t. <u>Gametes</u> – Haploid cells (n) produced by diploid cells (2n) for the purpose of sexual reproduction

u. <u>Virus</u> – A non-cellular infectious agent that has two characteristics:
        (1) It has genetic material (RNA or DNA) inside a protective protein coat.
        (2) It cannot reproduce on its own.

v. <u>Antibodies</u> – Specialized proteins that aid in destroying infectious agents

w. <u>Vaccine</u> – A weakened or inactive version of a pathogen that stimulates the body's production of antibodies which can aid in destroying the pathogen

2. Guanine and cytosine can bond together, as can adenine and thymine. In RNA, however, uracil replaces thymine. Thus when DNA has an adenine, RNA will have a uracil. When DNA has a thymine, RNA will have an adenine. When DNA has a cytosine, RNA will have a guanine, and when DNA has a guanine, RNA will have a cytosine. This makes the mRNA sequence:

a. <u>cytosine, guanine, uracil, uracil, adenine, cytosine</u>

b. It takes three nucleotide bases to code for an amino acid. Since this has six, it will code for <u>two amino acids</u>.

c. Each codon codes for one amino acid. Thus, there are <u>two codons</u> on the mRNA.

d. When mRNA has an adenine, tRNA will have a uracil. When mRNA has a uracil, tRNA will have an adenine. When mRNA has a cytosine, tRNA will have a guanine, and when mRNA has a guanine, tRNA will have a cytosine. This makes the tRNA anticodons' sequences:

<u>guanine, cytosine, adenine</u> and <u>adenine, uracil, guanine</u>

3. a. <u>transcription</u>   b. <u>translation</u>

4. This is <u>tRNA</u>, because only tRNA has anticodons.

5. If it is occurring in the ribosome, the protein is actually being assembled. This is <u>translation</u>.

6. <u>This would *not* mean that murders have no fault for what they do. Most genes only establish genetic trends. Environmental and spiritual factors affect the extent to which you follow those trends. Even if you have a genetic tendency to murder, the choices that you make can keep you from following that tendency.</u>

7. <u>It is not in interphase. Chromosomes only pack into their condensed form during reproduction.</u>

8. <u>Prophase, metaphase, anaphase, telophase</u>

9. a. Notice how there are two distinct nuclei far apart from each other and the plasma membrane is beginning to constrict. This is <u>telophase</u>.

b. The chromosomes are still in the nucleus, but they are distinct. This means that they are ready to start mitosis. Thus, this is <u>prophase</u>.

c. The chromosomes are lined up on the equatorial plane. This is <u>metaphase</u>.

d. The chromosomes are pulling away from each other, but they are not far apart. Also, the plasma membrane has not started to constrict. This is <u>anaphase</u>.

10. Diploid number is the total number of chromosomes in the cell. Haploid number is the number of homologous pairs. If there are a total of 16 chromosomes, then there must be 8 pairs. The haploid number is 8.

11. Since haploid number is the number of pairs, that tells us there are nine pairs. The diploid number is the total number of chromosomes in a diploid cell, which has *both members of each pair*. Since there are nine pairs, the diploid number is 9x2, or 18.

12. A gamete is haploid while a regular animal cell is diploid. This means that a gamete has only one chromosome from each homologous pair. A regular cell always both members of each homologous pair.

13. prophase I, metaphase I, anaphase I, telophase I, prophase II, metaphase II, anaphase II, telophase II.

14. Meiosis II: It is essentially mitosis acting on two haploid cells.

15. In meiosis I, a single diploid cell splits into two haploid cells with duplicated chromosomes. Thus, there are two cells. Since they are haploid, they have one chromosome from each pair. Since there are seven pairs, each cell has seven chromosomes. The chromosomes are duplicated, because the purpose of meiosis II is to separate the duplicates from the originals.

16. Before any meiosis started, these cells had seven pairs of chromosomes. When they went through meiosis I, they became haploid, so they now have seven chromosomes in total. The chromosomes are duplicated. In meiosis II, the duplicate chromosomes are separated from the originals, producing haploid cells with no duplicated chromosomes. Since there are four cells going through meiosis II, there are eight cells produced, there are seven chromosomes in each, but the chromosomes are not duplicated.

17. Male gametes are called sperm, while female gametes are called eggs.

18. Male animals produce four useful gametes with each meiosis, while female animals produce only one.

19. A polar body is a non-functional female gamete, because it is far too small to function properly. An egg is the one female gamete produced by meiosis that is large enough to function properly.

20. Sperm have flagella; thus, the male gamete can move on its own.

21. The lytic pathway is the way in which viruses reproduce, killing the cells of its host.

22. No virus is alive, because a virus cannot reproduce on its own.

23. A vaccine is only good if you take it before getting infected, because it is meant to build up the antibodies that you need to fight the virus off before it overwhelms your body.

SOLUTIONS TO THE STUDY GUIDE FOR MODULE #8

1. a. <u>True breeding</u> – If an organism has a certain characteristic that is always passed on to its offspring, we say that this organism bred true with respect to that characteristic.

b. <u>Allele</u> – One of a pair of genes that occupies the same position on homologous chromosomes

c. <u>Genotype</u> – Two-letter set that represents the alleles an organism possesses for a certain trait

d. <u>Phenotype</u> – The observable expression of an organism's genes

e. <u>Homozygous genotype</u> – A genotype in which both alleles are identical

f. <u>Heterozygous genotype</u> – A genotype with two different alleles

g. <u>Dominant allele</u> – An allele that will determine phenotype if just one is present in the genotype

h. <u>Recessive allele</u> – An allele that will not determine the phenotype unless the genotype is homozygous in that allele

i. Mendel's principles of genetics using updated terminology:

    1. <u>The traits of an organism are determined by its genes.</u>

    2. <u>Each organism has two alleles that make up the genotype for a given trait.</u>

    3. <u>In sexual reproduction, each parent contributes ONLY ONE of its alleles to the offspring.</u>

    4. <u>In each genotype, there is a dominant allele. If it exists in an organism, the phenotype is determined by that allele.</u>

j. <u>Pedigree</u> – A diagram that follows a particular phenotype through several generations

k. <u>Monohybrid cross</u> – A cross between two individuals, concentrating on only one definable trait

l. <u>Dihybrid cross</u> – A cross between two individuals, concentrating on two definable traits

m. <u>Autosomes</u> – Chromosomes that do not determine the sex of an individual

n. <u>Sex chromosomes</u> – Chromosomes that determine the sex of an individual

o. <u>Antigen</u> – A protein that, when introduced in the blood, triggers the production of an antibody

p. <u>Autosomal inheritance</u> – Inheritance of a genetic trait not on a sex chromosome

q. <u>Genetic disease carrier</u> – A person who is heterozygous in a recessive genetic disorder

r. <u>Sex-linked inheritance</u> – Inheritance of a genetic trait located on the sex chromosomes

s. <u>Mutation</u> – A radical chemical change in one or more alleles

t. <u>Change in chromosome structure</u> – A situation in which a chromosome loses or gains genes during meiosis

u. <u>Change in chromosome number</u> – A situation in which abnormal cellular events in meiosis lead to either none of a particular chromosome in the gamete or more than one chromosome in the gamete

2. a. <u>This homozygous genotype is "YY," resulting in a phenotype of yellow peas.</u>
    b. <u>This heterozygous genotype is "Yy," resulting in a phenotype of yellow peas.</u>
    c. <u>This homozygous genotype is "yy," resulting in a phenotype of green peas.</u>

3. <u>Meiosis separates the two alleles.</u>

4. One parent is homozygous dominant, so its genotype is "AA." The other is heterozygous, so its genotype is "Aa." The Punnett square looks like:

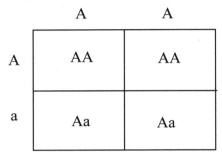

Thus, <u>50% of the offspring have the "AA" genotype and 50% have the "Aa" genotype.</u> Since each offspring has at least one of the dominant allele, however, <u>100% have the axial flower phenotype.</u>

5. Since the woman is heterozygous, her genotype is "Rr." The man cannot roll his tongue. Since the inability to roll your tongue is recessive, his genotype must be "rr." The resulting Punnett square is:

	R	r
r	Rr	rr
r	Rr	rr

Since having even one dominant allele allows you to be able to roll your tongue, <u>50% of the children will be able to roll their tongues.</u>

6. Since the square that represents the male parent is filled, it means that he has a black coat. This means he has at least one dominant allele. One of the offspring is white-coated. The only way that can happen is for each parent to have at least one recessive allele. Thus, <u>the genotype is "Bb."</u>

7. Individuals 1 and 2 can tell us which allele is dominant. After all, the offspring have both phenotypes. This means that at least one of them is homozygous recessive. Thus, each parent must

have the recessive allele. They both have no wings, but they must also carry the allele for wings, since one of their offspring has wings. This means no wings ("N") is the dominant allele. Since they each must also have the recessive allele, 1 and 2 must have the "Nn" genotype. Since some of the offspring between 3 and 4 also have recessive traits, they both must have the recessive allele. However, they do not express the recessive trait, so 3 and 4 must also have the "Nn" genotype.

8. Since the parent with smooth, yellow peas is homozygous, its genotype is "SSYY." Since the other expresses both recessive alleles, it must be homozygous in the recessive alleles. Thus, its genotype is "ssyy." Both of these parents can only produce one type of gamete each. The one parent can only produce a *SY* allele and the other can only produce a *sy*. This gives us a 1x1 Punnett square.

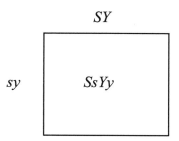

Since there is only one possible genotype, 100% of the offspring have the "SsYy" genotype and the smooth, yellow phenotype.

9. Since the parents are both heterozygous in each allele, their genotypes are "SsYy." There are 4 possible gametes: *SY, Sy, sY, sy*. The resulting Punnett square, then, is:

	SY	*Sy*	*sY*	*sy*
SY	*SSYY*	*SSYy*	*SsYY*	*SsYy*
Sy	*SSYy*	*SSyy*	*SsYy*	*Ssyy*
sY	*SsYY*	*SsYy*	*ssYY*	*ssYy*
sy	*SsYy*	*Ssyy*	*ssYy*	*ssyy*

smooth, yellow peas (genotypes SSYY, SsYy, SSYy, SsYY) 9 of 16 or 56.25 %
smooth, green peas (genotypes SSyy, Ssyy) 3 of 16 or 18.75 %
wrinkled, yellow peas (genotypes ssYY, ssYy) 3 of 16 or 18.75 %
wrinkled, green peas (genotype ssyy) 1 of 16 or 6.25 %

10. If the female is heterozygous, then her genotype is X^RX^r. Since the male is white-eyed, his genotype is X^rY. The resulting Punnett Square is:

	X^r	Y
X^R	X^RX^r	X^RY
X^r	X^rX^r	X^rY

Thus, 50% of the females (remember, only XX's are females) will be white-eyed and 50% of the males (only XY's are males) will be white-eyed.

11. If the male were white-eyed, then the Punnett square would look like the one above, resulting in 50% of the females having white eyes. If the male were red-eyed, however, the resulting Punnett square looks like this:

	X^R	Y
X^R	X^RX^R	X^RY
X^r	X^RX^r	X^rY

This Punnett square tells us that no white-eyed females (XX) are produced. Thus, the male's genotype is $\underline{X^RY}$.

12. If a gamete has two alleles for the same trait, it must have two of the same chromosome. In the fertilization process, then, there will be three chromosomes. Thus, a genetic disorder from a change in chromosome number will result.

13. The genetic disorder must be recessive. Thus, the person can carry the trait but, as long as he has the dominant allele, the person will not have the disease.

14. Sex-linked disorders affect men more frequently than women. This is because men have only one allele in sex-linked traits.

15. Not all traits are determined completely by genetics. Most are also determined by environmental factors and (in the case of humans) spiritual factors. While the genetics are the same, the environmental and spiritual factors were probably different.

16. Since the woman is type O, her genotype must be OO, as that would be the only way the recessive allele could be expressed. The man is type AB, so his genotype is "AB." The Punnett square, then, is given below:

	A	B
O	AO	BO
O	AO	BO

Since the O allele is recessive, the possible blood types for the children are A (50%) and B (50%).

17. Since the person is type B, the genotype must be either BB or BO. For the Rh-factor, the person expresses the recessive allele. Thus, her genotype must be homozygous in the recessive allele, which we called "pp" in "On Your Own" problem 8.10.

18. If a genetic trait is governed by many genes, we call it polygenetic inheritance.

ANSWERS TO EXPERIMENT 8.1

The free earlobe is dominant. If your pedigree can't determine this, that's fine. Now that you know the free earlobe is dominant, go back and review your pedigree to see if it makes sense. Here is an example of a family pedigree with respect to earlobe, using the filled circles and squares to represent free earlobes.

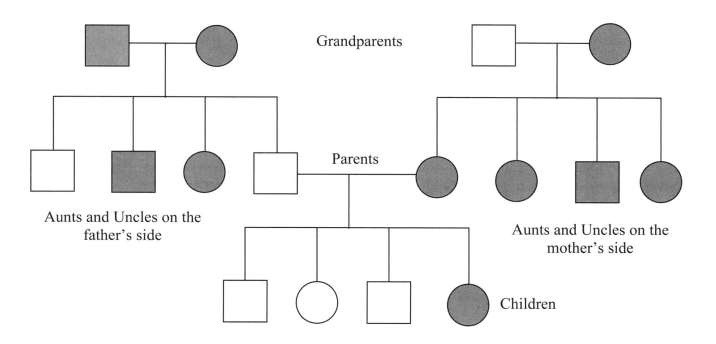

In this pedigree, you can see that the paternal grandparents have free earlobes, but they produce children with attached earlobes. Thus, they must each carry the recessive allele, so they are each heterozygous. The maternal grandfather has attached earlobes, and the maternal grandmother has free earlobes. Since all of their children have free earlobes, the maternal grandmother is probably homozygous in the free earlobe trait, because none of her children got an attached earlobe trait from her. The maternal grandfather must be homozygous in the attached earlobe trait, as that's the only way it could be expressed. In the same way, the father has attached earlobes and is therefore homozygous in that trait. The mother must be heterozygous, because she expresses the dominant trait, but since most of her children express the recessive trait, they had to get a recessive allele from her.

ANSWERS TO "EXPERIMENT" 8.2

1&2

	TR
tr	TtRr

3. <u>The genotype is TtRr, making all children with the taster/roller phenotype</u>.

4&5.

	TR	*Tr*	*tR*	*tr*
TR	*TTRR*	*TTRr*	*TtRR*	*TtRr*
Tr	*TTRr*	*TTrr*	*TtRr*	*Ttrr*
tR	*TtRR*	*TtRr*	*ttRR*	*ttRr*
tr	*TtRr*	*Ttrr*	*ttRr*	*ttrr*

6. roller / taster (genotypes TTRR, TtRr, TTRr, TtRR) 9 of 16 or <u>56.25 %</u>
   roller / nontaster (genotypes ttRR, ttRr) 3 of 16 or <u>18.75 %</u>
   nonroller / taster (genotypes TTrr, Ttrr) 3 of 16 or <u>18.75 %</u>
   nonroller / nontaster (genotype ttrr) <u>1 of 16 or 6.25 %</u>

ANSWERS TO "EXPERIMENT" 8.3

3.

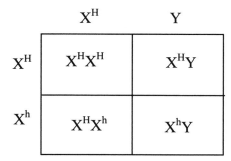

<u>The girls (XX) will never be hemophiliacs, but half of the boys (XY) will be.</u>

4. To have the disease, you must have only the recessive allele(s). For a son, that means <u>X^hY</u>

5. For a female to have the disease, she must have the recessive allele on both X chromosomes. Since one X always comes from the father, the father must have a recessive allele on his X chromosome. Also, the mother needs to have at least one, but she could have 2 recessive alleles. In the end, then, there are two possibilities, either of which is correct:

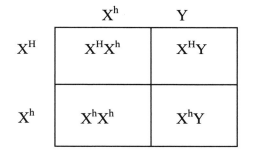   OR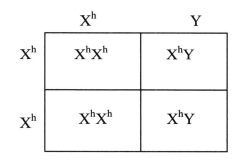

50% of the girls and 50% of the boys will have hemophilia.

All children will have hemophilia.

SOLUTIONS TO THE STUDY GUIDE FOR MODULE #9

1. a. <u>The immutability of species</u> – The idea that each individual species on the planet was specially created by God and could never fundamentally change

b. <u>Microevolution</u> – The theory that natural selection can, over time, take an organism and transform it into a more specialized species of that organism.

c. <u>Macroevolution</u> – The hypothesis that processes similar to those at work in microevolution can, over eons of time, transform an organism into a completely different kind of organism

d. <u>Strata</u> – Distinct layers of rock

e. <u>Fossils</u> – Preserved remains of once-living organisms

f. <u>Paleontology</u> – The study of fossils

g. <u>Structural homology</u> – The study of similar structures in different species

2. <u>He did most of his research while he was on board the HMS Beagle.</u> True, it took him years after leaving the Beagle before publishing, but that was mostly because of his wife's urgings not to publish. Although he made most of his observations that led to his theory on the Galapagos archipelago, it was on the ship that he did most of the work.

3. <u>No</u>. Stories like that are not true.

4. <u>Malthus believed in a constant struggle for survival</u>. Without this idea of a constant struggle, Darwin would have never come up with the concept of natural selection.

5. <u>Lyell came up with the idea that the present is the key to the past. He thought that the entire geological column could be explained by referring to the same processes that we see happening today</u>. Darwin basically took that same idea and applied it to his hypothesis. He said that the variation we see in nature is the result of the variations that occur in reproduction (which we see today) operating over eons of time.

6. <u>Darwin dispelled the idea of the immutability of the species</u>. By showing the evidence for microevolution, Darwin was able to show that species did change.

7. To go from a horse to a giraffe, there would need to be a lot added to the genetic code. Thus, this scenario is an example of <u>macroevolution</u>.

8. The fish remain fish; they have just varied their phenotype. Thus, this is variation within the genetic code, which is an example of <u>microevolution</u>.

9. <u>In microevolution, the same genetic code exists throughout the change. The changes that occur are simply the result of variation within that genetic code. In order for macroevolution to occur, information must be added to the genetic code, essentially creating a new genetic code</u>.

10. The data sets and their relation to macroevolution are given in the table below:

Data Set	Summary
The geological column	This data is inconclusive as far as macroevolution is concerned. If you believe that the geological column was formed according to the speculations of Lyell, it is evidence for macroevolution because it shows that life forms early in earth's history were simple and gradually got more complex. If you believe that the geological column was formed by natural catastrophe, then it is evidence against macroevolution. Since geologists have seen rock strata formed each way, it is impossible to tell which belief is scientifically correct.
The fossil record	This data is strong evidence against macroevolution. There are no clear intermediate links in the fossil record. The very few that macroevolutionists can produce are so similar to one of the two species they supposedly link, it is more scientifically sound to consider them a part of that species.
Structural homology	This data is strong evidence against macroevolution. The similar structures are not a result of inheritance from a common ancestor, because the similar structures are determined by quite different genes.
Molecular biology	This data is strong evidence against macroevolution. The vast majority of the data show no evolutionary patterns in the sequences of amino acids of common proteins.

11. *Australopithecus afarensis* is supposed to be an intermediate link between man and ape. However, every bone that we have found of this creature indicates it is an ape. Thus, it is safest to assume that it is an ape. *Archaeopteryx* is supposed to link birds and reptiles, but once again the fossils tell us it is just a bird.

12. The Cambrian Explosion refers to the fact that every major animal phylum in creation can be found in Cambrian rock. Thus, it is like there was an "explosion" of life. It presents two problems for macroevolution: (1) There is no way macroevolutionists can understand how macroevolution proceeded so quickly during those times. (2) There are just no intermediate links. In some parts of the geological column, you can find highly-questionable intermediate links, but they are at least something. In Cambrian rock, it just looks like the fossils appeared suddenly.

13. A bacterium can become resistant to antibiotics by conjugation, transformation, transduction, or mutation.

14. No information is added. In fact, in those cases studied, information is destroyed, leading to non-working or less efficient systems that just happen to make the bacterium resistant.

15. The most similar protein will be the one with the fewest difference in sequence. The protein in (a) has 5 amino acids different from the protein of interest, the protein in (b) has 4 differences and the one in (c) has 3. Thus, the protein in (c) is most similar.

16. A bacterium's cytochrome C should resemble a yeast's more than a kangaroo's does, because, according to evolutionists, the yeast evolved rather early after the bacterium, but the kangaroo came much, much later. In fact, however, the bacterium's cytochrome C sequence is *more similar to the kangaroo's* than it is to the yeast's!

17. <u>Neo-Darwinism hoped to provide a mechanism by which information could be added to the genetic code of an organism.</u> This was something Darwin's original hypothesis could not do.

18. <u>Punctuated equilibrium attempts to explain away the fact that the fossil record is devoid of any real intermediate links.</u>

19. <u>He would say that since the transition from species to species takes such a short amount of time, there is virtually no chance of an intermediate link being fossilized.</u>

20. <u>Structural homology and molecular biology still say that macroevolution (even by punctuated equilibrium) could not have happened.</u>

SOLUTIONS TO THE STUDY GUIDE FOR MODULE #10

1. a. <u>Ecology</u> – The study of the interactions between living and nonliving things

b. <u>Population</u> – A group of interbreeding organisms coexisting together

c. <u>Community</u> – A group of populations living and interacting in the same area

d. <u>Ecosystem</u> – An association of living organisms and their physical environment

e. <u>Biome</u> – A group of ecosystems classified by climate and plant life

f. <u>Primary consumer</u> – An organism that eats producers

g. <u>Secondary consumer</u> – An organism that eats primary consumers

h. <u>Tertiary consumer</u> – An organism that eats secondary consumers

i. <u>Ecological pyramid</u> – A diagram that shows the biomass of organisms at each trophic level

j. <u>Biomass</u> – A measure of the total dry mass of organisms within a particular region

k. <u>Transpiration</u> – Evaporation of water from the leaves of a plant

l. <u>Watershed</u> – An ecosystem where all water runoff drains into a single body of water

m. <u>Greenhouse effect</u> – The process by which certain gases (principally water vapor, carbon dioxide, and methane) trap heat that would otherwise escape the earth and radiate into space

2. <u>If an insect not native to the U.S. were carried into the country through foreign fruits and vegetables, it could ruin the balance of the U.S. ecosystem.</u>

3.

Organism	Possible Trophic Levels
Whale	primary consumer, secondary consumer
Sea turtle	primary consumer, secondary consumer
Phytoplankton	producer
Meran	primary consumer, secondary consumer
Ocean perch	secondary consumer
Zooplankton	primary consumer
Sea bass	secondary consumer (the meran can be a primary consumer, so if the sea bass eats a meran that eats only phytoplankton, it is a secondary consumer), tertiary consumer
Shark	secondary consumer, tertiary consumer

4. a. The size of the rectangle indicates biomass. Also, the rectangles, in order, represent producers (bottom), then primary consumers, then secondary consumers, then tertiary consumers (top). Finally, we must look at percentage change, not absolute changes in the length. Thus, <u>the primary and secondary consumers have the greatest disparity in biomass.</u>

b. The smallest amount of energy is wasted where the biomass is as close to equal as possible. Thus, <u>from producer to primary consumer wastes the least energy.</u>

5. <u>The clownfish and the sea anemone form a mutualistic symbiotic relationship. The clownfish is protected by the sea anemone and it attracts food to the sea anemone. The goby and the blind shrimp have a mutualistic symbiotic relationship in which the goby protects the blind shrimp, and the blind shrimp provides a home for the goby. Finally, the Oriental sweetlips and blue-streak wrasse form a mutualistic symbiotic relationship in which the sweetlips gets it teeth cleaned by the wrasse and the wrasse gets food from the sweetlips' teeth.</u>

6. <u>Mutualism seems to contradict the idea that organisms always battle for survival.</u>

7. <u>The ocean does not lose water because the land gets the excess water, and that excess water flows back into the ocean via surface runoff</u> (or as runoff from a river or stream).

8. <u>It transports nutrients</u> within an ecosystem and even from one ecosystem to another.

9. If too many trees and plants are removed from a watershed, <u>too many nutrients will flow into the river or stream, throwing off the ecosystem.</u>

10. <u>Oxygen is taken from the air principally by respiration and is restored principally by photosynthesis.</u>

11. <u>Oxygen is also removed from the air by fire, ozone formation, and the rusting of metals and minerals.</u>

12. <u>Oxygen is also restored by ozone destruction and water vapor destruction.</u>

13. <u>Carbon dioxide leaves the air by photosynthesis and by dissolving in the ocean.</u>

14. <u>Carbon dioxide enters the air via decomposition, fossil fuel burning, fire, and respiration.</u>

15. <u>Fuel burning</u> worries those who think that global warming is a problem, because it is a human-made way of adding more carbon dioxide to the air.

16. <u>No,</u> all measurable data indicate that any warming which did take place occurred before humans really started burning fuels in earnest.

17. <u>Nitrogen fixation is the process by which nitrogen gas from the atmosphere is converted into nitrogen-containing molecules that are useful to most of organisms in creation. Nitrogen-fixing bacteria perform it.</u>

18. <u>Organisms emit some nitrogen in their wastes, and the rest is turned back into useful forms of nitrogen by the decomposers that feed on their decaying remains.</u>

SOLUTIONS TO THE STUDY GUIDE FOR MODULE #11

1. a. <u>Invertebrates</u> – Animals that lack a backbone

b. <u>Vertebrates</u> – Animals that possess a backbone

c. <u>Spherical symmetry</u> – An organism possesses spherical symmetry if it can be cut into two identical halves by any cut that runs through the organism's center.

d. <u>Radial symmetry</u> – An organism possesses radial symmetry if it can be cut into two identical halves by any longitudinal cut through its center.

e. <u>Bilateral symmetry</u> – An organism possesses bilateral symmetry if it can only be cut into two identical halves by a single longitudinal cut along its center which divides it into right and left halves.

f. <u>Epidermis</u> – An outer layer of cells designed to provide protection

g. <u>Mesenchyme</u> – The jellylike substance that separates the epidermis from the inner cells in a sponge

h. <u>Collar cells</u> – Flagellated cells that push water through a sponge

i. <u>Amoebocytes</u> – Cells that move using pseudopods and perform a variety of functions in animals

j. <u>Gemmule</u> – A cluster of cells encased in a hard, spicule-reinforced shell

k. <u>Polyp</u> – The sessile, tubular form of a cnidarian with a mouth and tentacles at one end and a basal disk at the other

l. <u>Medusa</u> – A free-swimming cnidarian with a bell-shaped body and tentacles

m. <u>Epithelium</u> – Animal tissue consisting of one or more layers of cells that have only one free surface, because the other surface adheres to a membrane or other substance

n. <u>Mesoglea</u> – The jelly-like substance that separates the epithelial cells in a cnidarian

o. <u>Nematocysts</u> – Small capsules that contain a toxin which is injected into prey or predators

p. <u>Testes</u> – Organs that produce sperm

q. <u>Ovaries</u> – Organs that produce eggs

r. <u>Anterior end</u> – The end of an animal that contains its head

s. <u>Posterior end</u> – The end of an animal that contains its tail

t. <u>Circulatory system</u> – A system designed to transport food and other necessary substances throughout a creature's body

u. <u>Nervous system</u> – A system of sensitive cells that respond to stimuli such as sound, touch, and taste

v. <u>Ganglia</u> – Masses of nerve cell bodies

w. <u>Hermaphroditic</u> – Possessing both the male and the female reproductive organs

x. <u>Regeneration</u> – The ability to regrow a missing part of the body

y. <u>Mantle</u> – A sheath of tissue that encloses the vital organs of a mollusk, secretes its shell, and performs respiration

z. <u>Shell</u> – A tough, multilayered structure secreted by the mantle, generally used for protection, but sometimes for body support

aa. <u>Visceral hump</u> – A hump that contains a mollusk's heart, digestive, and excretory organs

bb. <u>Foot</u> – A muscular organ that is used for locomotion and takes a variety of forms depending on the animal

cc. <u>Radula</u> – An organ covered with teeth that mollusks use to scrape food into their mouths

dd. <u>Univalve</u> – An organism with a single shell

ee. <u>Bivalve</u> – An organism with two shells

2. <u>No</u>. All but one of the phyla in the animal kingdom are invertebrates (organisms with no backbones).

3. a. <u>Bilateral</u>, because it can only be cut into identical right and left halves

b. <u>Radial</u>, because any up and down cut through the center makes two identical halves

c. <u>Bilateral</u>, because it can only be cut into identical right and left halves

d. <u>Radial</u>, because any up and down cut through the center makes two identical halves

4. <u>Sponges get their prey by pulling water into themselves</u>. The water brings algae, bacteria, and organic matter that sponges eat.

5. <u>It contains spongin</u>, because spongin is soft. Spicules make a sponge hard and prickly. <u>These substances support the sponge.</u>

6. When asexually reproducing, sponges use <u>budding</u>.

7. <u>Amebocytes help digest and transport nutrients, they help carry waste to be excreted, they bring necessary gases such as oxygen to the cells, and they form the spicules or spongin.</u>

8. <u>A sponge produces gemmules during inclement times.</u>

9. Hydra nematocysts are triggered with pressure, while the sea anemone's are triggered chemically.

10. Cnidarians do not need these systems because their body walls are so thin that gases diffuse right through them.

11. Jellyfish spend part of their lives as polyps and the other part as medusas.

12. It must be in medusa form, because jellyfish can only reproduce sexually in medusa form.

13. Large coral colonies are called coral reefs.

14. a. mouth b. ventral nerve cord c. seminal receptacles d. seminal vesicles e. ventral blood vessel f. nephridia with nephridiopores g. clitellum h. intestine i. dorsal blood vessel j. gizzard k. crop l. oviduct m. ovary n. esophagus o. aortic arches p. pharynx q. ganglia

15. Earthworms bring minerals up from the lower parts of the soil and mix them with the nutrients at the top of the soil, which makes the soil fertile for plants. Their tunnels also allow oxygen to travel to the roots of a plant more easily.

16. If the first earthworm feels slimy near the clitellum, this means that it is covered with a slime coat. Thus, the first one must have recently mated but not yet produced a cocoon.

17. The earthworm is hermaphroditic and the hydra can be as well. However, although a hydra can sometimes mate with itself, an earthworm cannot.

18. The earthworm will suffocate, because oxygen cannot travel through a dry cuticle.

19. Planarians do not need circulatory systems because the intestine is so highly-branched that all cells are near it, so they can get their food directly from the intestine.

20. Without complex nervous or digestive systems, it must not need to seek out and fully digest prey. The only way it can survive, therefore, is by being parasitic.

21. When planarians asexually reproduce, they do so by regeneration.

22. a. Cnidaria   b. Mollusca   c. Porifera   d. Platyhelminthes   e. Annelida

SOLUTIONS TO THE STUDY GUIDE FOR MODULE #12

1. a. <u>Exoskeleton</u> – A body covering, typically made of chitin, that provides support and protection

b. <u>Molt</u> – To shed an old outer covering so that it can be replaced with a new one

c. <u>Thorax</u> – The body region between the head and the abdomen

d. <u>Abdomen</u> – The body region posterior to the thorax

e. <u>Cephalothorax</u> – A body region composed of the head and thorax fused together

f. <u>Compound eye</u> – An eye made of many lenses, each with a very limited scope

g. <u>Simple eye</u> – An eye with only one lens

h. <u>Open circulatory system</u> – A circulatory system that allows the blood to flow out of the blood vessels and into various body cavities so that the cells are in direct contact with the blood

i. <u>Statocyst</u> – The organ of balance in a crustacean

j. <u>Gonad</u> – A general term for the organ that produces gametes

k. <u>Complete metamorphosis</u> - Insect development consisting of four stages: egg, larva, pupa, and adult

l. <u>Incomplete metamorphosis</u> - Insect development consisting of three stages: egg, nymph, and adult

2. <u>Exoskeleton, body segmentation, jointed appendages, open circulatory system, and a ventral nervous system</u> are the common features of arthropods.

3. a. <u>antennae</u> b. <u>antennules</u> c. <u>cephalothorax</u> d. <u>abdomen</u> e. <u>telson</u> f. <u>uropods</u> g. <u>swimmerets</u> h. <u>carapace</u> i. <u>walking legs</u> j. <u>chelipeds</u>

4. a. <u>eye</u> b. <u>brain ganglia</u> c. <u>stomach</u> d. <u>gonad</u> e. <u>heart</u> f. <u>pericardial sinus</u> g. <u>intestine</u> h. <u>anus</u> i. <u>nerve cord</u> j. <u>digestive glands</u> k. <u>sternal sinus</u> l. <u>mouth</u> m. <u>esophagus</u> n. <u>green gland</u>

5. <u>Blood collects in the pericardial sinus, and it enters the heart through one of three openings in the heart's surface. Each opening has a valve that closes when the heart is ready to pump. Once it absorbs the blood and closes these valves, the heart pumps blood through a series of blood vessels that are open at the end. These vessels dump directly into various body cavities. Gravity causes the blood to fall into the sternal sinus, where it is collected by blood vessels that are open at one end. Unlike the blood vessels that dump the blood into the body cavities, these vessels carry the blood back towards the pericardial sinus. On its way there, the blood is passed through the gills where it can release the carbon dioxide it has collected and pick up a fresh supply of oxygen. The blood also passes through green glands, which clean it of impurities and dump those impurities back into the surroundings. Once the blood has passed through the gills and the green glands, it then makes its way back to the pericardial sinus to begin the trip all over again.</u>

6. <u>It cleans the blood of impurities.</u>

7. <u>The swimmerets and maxillae are important.</u> Without them, fresh, oxygen-rich water would not enter the gill chambers.

8. <u>The injury gets sealed off to prevent bleeding, and then a new limb regenerates.</u>

9. <u>They are attached to the swimmerets.</u>

10. <u>They molt because their exoskeletons get too small for their growing bodies.</u>

11. <u>The antennules and antennae are responsible for taste and touch.</u>

12. <u>Four pairs of walking legs, two segments in body, no antennae, book lungs, four pairs of simple eyes.</u>

13. Some species of spider build a <u>sheet web</u>, which is a single, flat sheet of sticky silk. Some spiders spin <u>tangle webs</u> that have no real discernible pattern. Some spin <u>orb webs</u> consisting of concentric circles of sticky silk that are supported by "spokes" of non-sticky silk.

14. <u>No</u>, some spiders spin silk to make trap doors, and some even fire their silk like a projectile.

15. <u>The lung has many thin layers that look like the pages of a book.</u>

16. <u>Three pairs of walking (or jumping) legs, wings, three segments in body, one pair of antennae.</u>

17. <u>Insects do not need respiratory systems because of a complex network of tracheas that allow air to travel throughout the body.</u>

18. The pupa stage only exists in <u>complete metamorphosis</u>.

19. <u>membranous wings, scaled wings, leather-like wings, and horny wings</u>.

20. a. <u>Orthoptera</u>

b. <u>Hymenoptera</u>

c. <u>Diptera</u>

d. <u>Coleoptera</u>

e. <u>Lepidoptera</u>

ANSWERS TO EXPERIMENT 12.2

a. If the armor-like wings and the war-machine appearance didn't give it away, the horn on the head should have. This is from <u>order Coleoptera</u>.

b. The two membranous wings with two smaller membranous wings (balancers) should be a hint. This is a mosquito from <u>order Diptera</u>.

c. The scaled wings tell you that this butterfly belongs in <u>order Lepidoptera</u>.

d. This is a katydid, which belongs to <u>order Orthoptera</u>. The wings appear to be either leathery or horned, you can't tell by the picture. However, the long hind legs should give it away.

e. The membranous wings could mean order Diptera, but the tapered posterior hints at a stinger. This is a wasp from <u>order Hymenoptera</u>.

f. Note the large mandibles and the exoskeleton covering the wings. These are from <u>order Coleoptera</u>.

SOLUTIONS TO THE STUDY GUIDE FOR MODULE #13

1. a. <u>Vertebrae</u> – Segments of bone or some other hard substance that are arranged into a backbone

b. <u>Notochord</u> – A rod of tough, flexible material that runs the length of a creature's body, providing the majority of its support

c. <u>Endoskeleton</u> – A skeleton on the inside of a creature's body, typically composed of bone or cartilage

d. <u>Bone marrow</u> – A soft tissue inside the bone that produces blood cells

e. <u>Axial skeleton</u> – The portion of the skeleton that supports and protects the head, neck, and trunk

f. <u>Appendicular skeleton</u> – The portion of the skeleton that attaches to the axial skeleton and has the limbs attached to it

g. <u>Closed circulatory system</u> – A circulatory system in which the oxygen-carrying blood cells never leave the blood vessels

h. <u>Arteries</u> – Blood vessels that carry blood away from the heart

i. <u>Capillaries</u> – Tiny, thin-walled blood vessels that allow the exchange of gases and nutrients between the blood and the cells of the body

j. <u>Veins</u> – Blood vessels that carry blood back to the heart

k. <u>Olfactory lobes</u> – The lobes of the brain that receive signals from the receptors in the nose

l. <u>Cerebrum</u> – The lobes of the brain that integrate sensory information and coordinate the creature's response to that information

m. <u>Optic lobes</u> – The lobes of the brain that receive signals from the receptors in the eyes

n. <u>Cerebellum</u> – The lobe that controls involuntary actions and refines muscle movement

o. <u>Medulla oblongata</u> – The lobes that coordinate vital functions, such as those of the circulatory and respiratory systems, and transport signals from the brain to the spinal cord

p. <u>Internal fertilization</u> – The process by which the male places sperm inside the female's body, where the eggs are fertilized

q. <u>External fertilization</u> – The process by which the female lays eggs and the male fertilizes them once they are outside of the female

r. <u>Oviparous development</u> – Development that occurs in an egg that is hatched outside the female's body

s.   <u>Ovoviviparous development</u> – Development that occurs in an egg that is hatched inside the female's body

t.   <u>Viviparous development</u> – Development that occurs inside the female, allowing the offspring to gain nutrients and vital substances from the mother through a placenta

u.   <u>Anadromous</u> – A life cycle in which creatures are hatched in fresh water, migrate to salt water as adults, and then go back to fresh water in order to reproduce

v.   <u>Bile</u> – A mixture of salts and phospholipids that aids in the breakdown of fat

w.   <u>Atrium</u> – A heart chamber that receives blood

x.   <u>Ventricle</u> – A heart chamber from which blood is pumped out

y.   <u>Ectothermic</u> – Lacking an internal mechanism for regulating body heat

z.   <u>Hibernation</u> – A state of extremely low metabolism and respiration, accompanied by lower-than-normal body temperatures

2. a. <u>Class Amphibia</u>   b. <u>Class Chondrichthyes</u>   c. <u>Subphylum Cephalochordata</u>   d. <u>Class Osteichthyes</u>   e. <u>Subphylum Urochordata</u>   f. <u>Class Agnatha</u>

3. In addition to the other common features of members of phylum Chordata, <u>they all go through metamorphosis from larva to adult.</u>

4. Bone is made of collagen fibers that have been hardened with calcium, while cartilage is not hardened. Thus, <u>cartilage is more flexible and weaker than bone.</u>

5. Capillaries have thin walls to allow for the diffusion of gases. Thus, <u>this is, most likely, a capillary.</u>

6. <u>Red blood cells carry oxygen in the blood.</u>

7. <u>Hemoglobin gives red blood cells their color.</u>

8. The cerebellum refines muscle movement. A creature that has uncoordinated, jerky muscle movements has a small cerebellum. Thus, an <u>amphibian has a small cerebellum.</u>

9. Vertebrates have enlarged lobes if the creature has a particular aptitude for the function controlled by the lobe. Since owls have good eyesight, <u>their optic lobes are enlarged.</u>

10. <u>Fertilization is internal,</u> because the female takes the sperm in to form the zygote, which is then encased in the egg. <u>Development is oviparous,</u> because the egg hatches externally.

11. The stronger the skeleton, the less flexible it is. Lampreys and rays both have cartilaginous skeletons, but the salmon is a bony fish. Thus, <u>the salmon's skeleton is less flexible.</u>

12. <u>Atlantic salmon and many lamprey are anadromous.</u>

13. The shark's most sensitive means of finding prey is its electrical field sensor.

14. The lateral line senses vibrations in the water. This alerts fish and sharks to movements in the water. Typically, sharks investigate the vibrations as possible food sources, while most bony fish swim away from them in fear.

15. In both sharks and bony fish, the dorsal fins are used for balance in the water. In bony fish, the anterior dorsal fin is also a defensive weapon, because it is hard and sharp.

16. Rays have thin, whiplike tails, while skates have thicker, fleshy tails.

17. a. esophagus b. brain c. spinal cord d. stomach e. air bladder f. kidney g. gonad h. anus
i. intestine j. pyloric ceca k. gall bladder l. liver m. heart n. gills

18.

Organ	Basic Function
Gills	Exchange of carbon dioxide and oxygen between the water and the blood
Heart	Pumps blood
Liver	Makes bile for the digestion of fats and does many other chemical tasks
Gall bladder	Concentrates bile
Pyloric ceca	Secretes digestive enzymes and chemicals that break down food in stomach
Intestine	Digests food
Gonad	Reproduction
Anus	Expelling of undigested food
Brain	Controls nervous system
Esophagus	Sends food to stomach
Stomach	Stores and breaks down food
Spinal cord	Sends messages from brain to other parts of the body and vice-versa
Kidney	Cleans blood of waste products
Air bladder	Allows fish to change depths and float in water

19. a. anterior cardial vein b. efferent brachial arteries c. dorsal aorta d. kidney
e. posterior cardial vein f. atrium g. ventricle h. ventral aorta i. afferent brachial arteries j. gills

20.

Arteries	Veins	Neither
Efferent brachial arteries	Anterior cardial vein	Atrium
Dorsal aorta	Posterior cardial vein	Ventricle
Ventral aorta		Gills
Afferent brachial arteries		Kidney

21. Their endoskeleton is made mostly of bone.
   Their skin is smooth with many capillaries and pigments. Amphibians do not have scales.
   They usually have two pairs of limbs with webbed feet.
   They have as many as four organs of respiration.
   They have a three-chambered heart.
   They are oviparous with external fertilization.

22. Frogs have smooth, wet skin and live near water. Toads have dry, warty skin and need not live near water.

23. The major respiratory organ for most amphibians is the skin.

SOLUTIONS TO THE STUDY GUIDE FOR MODULE #14

1. a. <u>Botany</u> – The study of plants

b. <u>Perennial plants</u> – Plants that grow year after year

c. <u>Annual plants</u> – Plants that live for only one year

d. <u>Biennial plants</u> – Plants that live for two years

e. <u>Vegetative organs</u> – The parts of a plant (such as stems, roots, and leaves) that are not involved in reproduction

f. <u>Reproductive plant organs</u> – The parts of a plant (such as flowers, fruits, and seeds) involved in reproduction

g. <u>Undifferentiated cells</u> – Cells that have not specialized in any particular function

h. <u>Xylem</u> – Nonliving vascular tissue that carries water and dissolved minerals from the roots of a plant to its leaves

i. <u>Phloem</u> – Living vascular tissue that carries sugar and organic substances throughout a plant

j. <u>Leaf mosaic</u> – The arrangement of leaves on the stem of a plant

k. <u>Leaf margin</u> – The characteristics of the leaf edge

l. <u>Deciduous plant</u> – A plant that loses its leaves for winter

m. <u>Girdling</u> – The process of cutting away a ring of inner and outer bark all the way around a tree trunk

n. <u>Alternation of generations</u> – A life cycle in which there is both a muticellular diploid form and a multicellular haploid form

o. <u>Dominant generation</u> – In alternation of generations, the generation that occupies the largest portion of the life cycle

p. <u>Pollen</u> – A fine dust that contains the sperm of seed-producing plants

q. <u>Cotyledon</u> – A "seed leaf" which develops as a part of the seed - it provides nutrients to the developing seedling and eventually becomes the first leaf of the plant.

2. <u>Meristematic tissue</u> will be anywhere that mitosis is going on. The cells that perform mitosis are a part of the meristematic tissue.

3. <u>The petiole</u> attaches the leaf blade to the stem.

4. a. <u>Whorled</u>   b. <u>Alternate</u>   c. <u>Opposite</u>

5.

Letter	Shape	Margin	Venation
a.	Deltoid	Entire	Parallel
b.	Elliptical	Serrate	Pinnate
c.	Lobed	Entire	Pinnate
d.	Cleft	Dentate	Palmate
e.	Orbicular	Undulate	Pinnate (This is a tough one. You might think it's parallel, but there is actually a vein in the middle, from which the other veins sprout.)
f.	Chordate	Entire	Pinnate

6. a. <u>photosynthesis</u> b. <u>photosynthesis</u> c. <u>protection</u> d. <u>transports water and minerals</u> e. <u>transports food and organic substances</u> f. <u>support</u>

7. <u>The guard cells control the opening and closing of the stomata.</u>

8. <u>The spongy mesophyll is typically on the underside of the leaf, and it is usually a lighter shade of green due to the fact that the photosynthesis cells are not as tightly packed there.</u>

9. <u>Carotenoids and anthocyanins.</u>

10. <u>No, a tree without an abscission layer cannot be deciduous.</u> Remember, the abscission layer cuts off the flow of nutrients to the leaves, which causes them to stop doing photosynthesis, causing them to die. With no abscission layer, that will not happen and the tree will not lose its leaves in the winter.

11. <u>The abscission layer is right between the stem and the petiole.</u>

12. The four regions of a root are: the <u>root cap, the meristematic region, the elongation region, and the maturation region. The undifferentiated cells are in the meristematic region.</u>

13. a. <u>This is from a dicot.</u> The fibrovascular bundles do not have a face-like appearance; instead, they are characteristic of dicots.

b. <u>This is from a monocot.</u> The face-like characteristic of the fibrovascular bundles tells you this.

14. <u>Woody stems have no limit to their growth because the cork cambium can always produce more bark.</u> Thus, when the bark cracks, the inner parts of the stem are not exposed to the surroundings.

15. <u>The vascular cambium produces new vascular tissue.</u>

16. <u>It is woody.</u> The cork cambium appears only in woody stems. It makes new cork tissue for the outer bark.

17. <u>Xylem make up most of the wood in a woody stem, while phloem are found in the inner bark.</u>

18. <u>The dominant generation in mosses is the gametophyte generation, and it is haploid.</u>

19. If it has archegonia and antheridia, it produces gametes. Thus, <u>it is in the gametophyte generation, which is not the dominant generation for ferns.</u>

20. <u>Since plants from phylum Bryophyta have no vascular tissue, there is no efficient way to transport nutrients throughout the plant.</u> The plant must therefore stay small so that the nutrients need not travel far.

21. <u>The plant must have a fibrous root system.</u> If a root system does not go deeper than the height of a plant, it must spread out so that its total length is greater than that of the plant.

22. <u>The female reproductive organ is the seed cone, and the male is the pollen cone.</u>

23. <u>The number of cotyledons produced in the seed is the fundamental difference between monocots and dicots.</u> Monocots have one cotyledon in their seeds, dicots have two.

24. <u>In monocots, the venation is parallel, while it is netted (pinnate or palmate) in dicots. The fibrovascular bundles are packaged differently in monocots and dicots. Typically, monocots have fibrous root systems whereas dicots have taproot systems. Finally, monocots usually produce flowers in groups of three or six while dicots produce flowers in groups of four or five.</u> The student need list only one of these.

25. <u>It belongs in phylum Coniferophyta</u>, which contains the cone-making plants. <u>It is vascular.</u> Only the bryophytes are nonvascular.

26. <u>It belongs in phylum Anthophyta</u>.

SOLUTIONS TO THE STUDY GUIDE FOR MODULE #15

1. a. <u>Physiology</u> – The study of life processes in an organism

b. <u>Nastic movement</u> – A plant's response to a stimulus such that the direction of the response is preprogrammed and not dependent of the direction of the stimulus

c. <u>Pore spaces</u> – Spaces in the soil that determine how much water and air the soil can hold

d. <u>Loam</u> – A mixture of gravel, sand, silt, clay, and organic matter

e. <u>Cohesion</u> – The phenomenon that occurs when individual molecules are so strongly attracted to each other that they tend to stay together, even when exposed to tension

f. <u>Translocation</u> – The process by which organic substances move through the phloem of a plant

g. <u>Hormones</u> – Chemicals that circulate throughout multicellular organisms, regulating cellular processes by interacting with specifically targeted cells

h. <u>Phototropism</u> – A growth response to light

i. <u>Gravitropism</u> – A growth response to gravity

j. <u>Thigmotropism</u> – A growth response to touch

k. <u>Perfect flowers</u> – Flowers with both stamens and carpels

l. <u>Imperfect flowers</u> – Flowers with either stamens or carpels, but not both

m. <u>Pollination</u> – The transfer of pollen grains from the anther to the carpel in flowering plants

n. <u>Double fertilization</u> – A fertilization process that requires two sperm to fuse with two other cells

o. <u>Seed</u> – An ovule with a protective coating, encasing a mature plant embryo and a nutrient source

p. <u>Fruit</u> – A mature ovary that contains a seed or seeds

2. <u>A plant uses water for photosynthesis, turgor pressure, hydrolysis, and transport.</u> Since a plant can wilt without dying, <u>turgor pressure can be ignored for a short time.</u>

3. <u>The first plant is using nastic movements and the second is using phototropism.</u> Nastic movements refer to movements that happen in a pre-programmed direction. Phototropism is directional, depending on the direction of the stimulus.

4. <u>The cohesion-tension theory states that when water evaporates through the stomata in a plant's leaves, a deficit of water is created. This causes the water molecules just below those that evaporated to move up and take their place. Since water molecules like to stay together, however, the water</u>

molecules just below the ones that moved up also move up, in order to stay close. This causes a chain reaction, eventually causing water from the roots to move up into other parts of the plant.

5. Xylem cells need not be alive for xylem to do their job. Since we think that the cohesion-tension theory of water explains how water and dissolved nutrients travel up a plant, the xylem cells need not play an active role in the transport.

6. Phloem cells must be alive in order for the phloem to do their job, because the phloem cells take an active part in translocation.

7. Xylem contain water and dissolved minerals, while phloem contain sugar and organic substances.

8. Insectivorous plants do not really eat insects. They decompose the insects and use their raw materials for biosynthesis. Insectivorous plants produce their own food via photosynthesis just like other plants.

9. Vegetative reproduction leads to offspring with genetic codes which are identical to the parent. Sexual reproduction leads to offspring with genetic codes which are similar to, but not identical to, the parents' genetic codes.

10. The gardener must have grafted limbs from a tree that produces normal-sized apples to his crabapple tree.

11. The stamen is the male reproductive organ, and the carpel is the female reproductive organ. The carpel is sometimes called the pistil.

12. Both structures are multicellular, and they both reproduce using gametes. This is the basic definition of the gametophyte generation in an alternations of generation life cycle.

13. There is at least one sperm cell, and there is a tube nucleus.

14. Typically, there are seven cells in an embryo sac. Remember, the megaspore undergoes mitosis three times to make eight nuclei. Then, the cell segments into six small cells and one large cell that has two nuclei. Two of these cells get fertilized. One becomes the zygote, and one becomes the endosperm.

15. a. stigma b. style c. ovary d. ovule e. sepal f. anther g. filament h. petal i. receptacle j. pedicel

16. Parts a, b, and c make up the carpel.

17. Parts f and g make up the stamen.

18. Pollination is simply the transfer of pollen from an anther to a stigma, while fertilization is the act of the sperms fusing with the egg and the large central nucleus in the embryo sac. You can use the terms "stamen" and "carpel" instead of "anther" and "stigma," but the latter are more precise.

19. Two sperms cells are used, because plants engage in double fertilization.

20.   The endosperm comes from the fertilization of the large, two-nucleus cell that is at the center of the embryo sac. It provides nutrition for the developing embryo.

21.   Cotyledons either absorb the endosperm or aid in the transfer of nutrients from the endosperm to the embryo. This is how cotyledons provide a plant with nutrition before germination. After germination, they often perform the first photosynthesis in the plant.

22.   The three basic parts are the radicle, the hypocotyl, and the epicotyl. The radicle becomes the root, the hypocotyl the stem, and the epicotyl gives rise to the first true leaves of the plant.

23.   A fruit allows for the dispersal of seeds to places away from the parent.

24.   There are many possible answers. The student needs at least three:

wind, bees, beetles, birds, moths, or butterflies

25. They form leaf-life structures if they end up rising above ground with the seedling. They often even carry out photosynthesis for a while.

SOLUTIONS TO THE STUDY GUIDE FOR MODULE #16

1. a. <u>Amniotic egg</u> – A shelled, water-retaining egg that allows reptile, bird, and certain mammal embryos to develop on land

b. <u>Neurotoxin</u> – A poison that attacks the nervous system, causing blindness, paralysis, or suffocation

c. <u>Hemotoxin</u> – A poison that attacks the red blood cells and blood vessels, destroying circulation

d. <u>Endotherm</u> – An organism that is internally warmed by a heat-generating metabolic process

e. <u>Down feathers</u> – Feathers with smooth barbules but no hooked barbules

f. <u>Contour feathers</u> – Feathers with hooked and smooth barbules, allowing the barbules to interlock

g. <u>Placenta</u> – A structure that allows an embryo to be nourished with the mother's blood supply

h. <u>Gestation</u> – The period of time during which an embryo develops before being born

i. <u>Mammary glands</u> – Specialized organs in mammals that produce milk to nourish the young

2. The common characteristics of reptiles are:

- <u>Covered with tough, dry scales</u>
- <u>Ectothermic</u>
- <u>Breathe with lungs throughout their lives</u>
- <u>Three-chambered heart with a ventricle that is partially divided</u>
- <u>Produce amniotic eggs covered with a leathery shell, most oviparous, some ovoviviparous</u>

3. <u>Reptiles are ectothermic, while birds and mammals are endothermic.</u>

4.   a. <u>amniotic fluid</u> b. <u>embryo</u> c. <u>amnion</u> d. <u>allantois</u> e. <u>chorion</u> f. <u>yolk sac</u> g. <u>yolk</u> h. <u>albumen</u> i. <u>shell</u>

5. <u>The yolk serves as nourishment for the developing embryo. The allantois allows the embryo to breathe, and the albumen destroys pathogens that can enter the egg as well as supplying water and amino acids to the embryo. The albumen also acts as a shock-absorber.</u>

6. <u>They must both molt because their body covering is not living.</u>

7. <u>Reptile scales prevent water loss and insulate the reptile's body.</u>

8. a. <u>Squamata</u>   b. <u>Rhynchocephalia</u>   c. <u>Squamata</u>   d. <u>Testudines</u> e. <u>Crocodilia</u>
     f. <u>Testudines</u>

9. The common characteristics are:

- <u>Endothermic</u>
- <u>Heart with four chambers</u>
- <u>Toothless bill</u>
- <u>Oviparous, laying an amniotic egg that is covered in a lime-containing shell</u>
- <u>Covered with feathers.</u>
- <u>Skeleton composed of porous, lightweight bones (not a characteristic for all birds)</u>

10. <u>No.</u> There are two orders of flightless birds.

11. <u>If the blood sample has a mixture of oxygen-rich and oxygen-poor blood, it comes from an amphibian. If it has only one or the other, it comes from a bird.</u> Remember, a bird's heart has four chambers, so oxygen-rich and oxygen-poor blood do not mix!

12. <u>A bird egg's shell is harder</u>, because it contains lime.

13. <u>Down feathers</u> have no hooked barbules.

14. <u>Contour feathers are used for flight, while down feathers are used for insulation.</u>

15. <u>When preening, a bird is actually oiling its feathers.</u> The feathers need to be oiled regularly to keep the hooked barbules sliding freely along the smooth barbules and to keep the feathers essentially waterproof.

16. <u>A bird's feathers molt in pairs</u>, one or a few at a time. This is different from arthropods and reptiles, which tend to molt their whole body covering at once.

17. <u>Flight engineers learned the proper structure of a wing from birds. They also learned how to make strong, hollow tubes from studying bird bones. Finally, they learned how to reduce wing turbulence from birds.</u>

18. <u>The amphibian's bone is heavier.</u> Birds have air-filled cavities that make their bones lighter than other vertebrates' bones.

19. The common characteristics are:

- <u>Hair covering the skin</u>
- <u>Reproduce with internal fertilization and usually viviparous</u>
- <u>Nourish their young with milk secreted from specialized glands</u>
- <u>Four-chambered heart</u>
- <u>Endothermic</u>

20. <u>Underhair's main job is insulation.</u>

21. <u>We usually see the mammal's guard hair</u>, because that's what's on top.

22. Any mammal from orders Monotremata or Marsupialia is non-placental. Thus, <u>any one of the following: duck-billed platypuses, spiny anteaters, kangaroos, wallabies, koalas, opossums</u>.

23. <u>Offspring born after a long gestation period are more developed than those born after a short gestation period</u>.

Answers to the Module Summaries in Appendix B

Not all of the blanks have to be filled in with exactly the phrases used here. As long as the general message of each paragraph is the same, that's fine.

ANSWERS TO THE SUMMARY OF MODULE #1

1. Four characteristics of life:

a. All life forms contain <u>deoxyribonucleic acid</u>, which is called <u>DNA</u>.
b. All life forms have a method by which they <u>extract energy</u> from the surroundings and convert it into <u>energy that sustains them</u>.
c. All life forms can <u>sense changes</u> in their surroundings and <u>respond to those changes</u>.
d. All life forms <u>reproduce</u>.

2. DNA provides the <u>information</u> necessary to take a bunch of lifeless chemicals and turn them into <u>an ordered, living system</u>.

3. <u>Metabolism</u> can be split into two categories: (1) <u>anabolism</u>, which involves using energy and simple chemical building blocks to produce large chemicals and structures and (2) catabolism, which involves <u>breaking down chemicals to produce energy and simple chemical building blocks</u>.

4. The vast majority of energy that sustains life comes from <u>the sun</u>. <u>Producers</u> use that energy to make food for themselves via a process called <u>photosynthesis</u>. Consumers get energy from the producers by <u>eating them</u>. Consumers can be split into three categories: <u>herbivores</u> (which eat only plants), <u>carnivores</u> (which eat only nonplants), and <u>omnivores</u> (which eat plants and nonplants). The energy of dead producers and consumers is recycled back into creation by the <u>decomposers</u>.

5. Producers are often called <u>autotrophs</u>, the Greek roots of which literally mean "self-feeder." Consumers and decomposers are often called <u>heterotrophs</u>, which literally means "<u>other-feeder</u>."

6. Living organisms are equipped with structures called <u>receptors</u>, which receive information about their surroundings. God's creation is always <u>changing</u>, which is why these structures are necessary for survival.

7. In asexual reproduction, the characteristics and traits inherited by the offspring are, under normal circumstances, <u>identical</u> to the parent. In sexual reproduction, under normal circumstances, the offspring's traits and characteristics are <u>some mixture of each parent's traits and characteristics</u>. When <u>mutations</u> occur, the offspring can possess traits that are incredibly different from those of the parent or parents.

8. In the scientific method, the scientist starts by <u>observing</u> the world around him. He then forms a <u>hypothesis</u> to explain some aspect of how the world functions. He then <u>collects data</u> in an attempt to test his <u>hypothesis</u>. If a large amount of <u>data</u> confirms the <u>hypothesis</u>, it becomes a <u>theory</u>, which is tested with even more <u>data</u>. If it continues to be confirmed over several generations, it might become a <u>scientific law</u>.

9. Scientists once believed that life could spring from nonliving things. This was called <u>spontaneous generation</u>, and it was refuted in the mid 1800s by <u>Louis Pasteur</u>. The story of how the scientific community believed in it for so long demonstrates that science has <u>limitations</u>.

10. The newest version of spontaneous generation is called <u>abiogenesis</u>, and it claims that long ago, <u>very simple life forms spontaneously appeared through chemical reactions</u>.

11. The groups used in our classification scheme, from largest to smallest are: <u>kingdom, phylum, class, order, family, genus</u>, and <u>species</u>.

12. The five kingdoms we use in this course are: <u>Monera, Protista, Fungi, Plantae</u>, and <u>Animalia</u>.

13. A cell with no membrane-bounded organelles is <u>prokaryotic</u>, while one with membrane-bounded organelles is a <u>eukaryotic</u>. Members of kingdom Monera are composed of <u>prokaryotic cells</u>.

14. A unit of one or more populations of individuals that can reproduce under normal conditions, produce fertile offspring, and are reproductively isolated from other such units is called a <u>species</u>.

15. A series of questions that is designed to classify organisms is called a biological <u>key</u>.

16. When we call wolves *"Canis lupus,"* we are using <u>binomial nomenclature</u>.

17. In the <u>three-domain</u> system of classification, the three basic groups are <u>Archaea, Bacteria</u>, and <u>Eukarya</u>. Members of kingdom Monera are placed in either <u>Archaea</u> or <u>Bacteria</u>, and all of the other kingdoms are placed in <u>Eukarya</u>.

18. A creationist taxonomy scheme that attempts to classify organisms based on the kind of organisms that God made during creation is called <u>baraminology</u>.

19. Multicellular autotrophs are typically placed in kingdom <u>Plantae</u>.

20. Single-celled creatures made of eukaryotic cells are placed in kingdom <u>Protista</u>.

21. Multicellular consumers are typically placed in kingdom <u>Animalia</u>.

22. Decomposers made of eukaryotic cells are mostly found in kingdom <u>Fungi</u>.

23. Organisms made of prokaryotic cells are found in kingdom <u>Monera</u>.

ANSWERS TO THE SUMMARY OF MODULE #2

1. The term <u>bacteria</u> is often used as a general term that applies to all members of kingdom Monera. Some are beneficial to humans, but some are <u>pathogenic</u>, which means they cause disease.

2. Some bacteria have a <u>capsule</u> that surrounds the cell wall. It is composed of an organized layer of sticky sugars that <u>help the bacteria adhere to surfaces</u>. It is also a protective layer that <u>deters infection-fighting agents</u>.

3. Most bacteria have a <u>cell wall</u> that holds the contents of the bacterium together, regulates the amount of water that a bacterium can absorb, and holds the cell into one of three shapes: <u>spherical (coccus), rod-shaped (bacillus), or helical (spirillum)</u>. The absence or presence of a <u>cell wall</u> as well as its composition (if it exists) are used to <u>classify</u> bacteria.

4. Underneath the cell wall (if it exists), there is a <u>plasma membrane</u>, which regulates what the bacterium takes in from the outside world.

5. <u>Cytoplasm</u> exists throughout the interior of a cell, supporting the DNA and the ribosomes.

6. Many bacteria have fibrous bristles called <u>fimbriae</u>, which are used for grasping. Locomotion is accomplished with a <u>flagellum</u>.

7. Ribosomes make special chemicals known as <u>proteins</u>.

8. In terms of what they eat, most bacteria are <u>decomposers</u>. As a result, they are called <u>saprophytes</u>.

9. Some bacteria are <u>parasites</u>, which means they feed on a living host.

10. Autotrophic bacteria manufacture their own food by either <u>photosynthesis</u> (using the energy of the sun to make food) or <u>chemosynthesis</u> (promoting chemical reactions that release energy).

11. Some bacteria are <u>aerobic</u>, which means they need oxygen in order to survive. Some are <u>anaerobic</u>, which means they do not need oxygen. The latter bacteria either <u>decompose</u> dead organisms or <u>convert useless chemicals</u> into chemicals that can be used by other life forms.

12. Asexual reproduction in bacteria is often called <u>binary fission</u>.

13. Typically, a population of bacteria starts off with <u>exponential</u> growth until it reaches a <u>steady state</u> in which bacteria die as quickly as new ones are made. When population growth is controlled by resources, we call it <u>logistic growth</u>. If resources begin to run out, the population will <u>decrease</u>.

14. When bacteria exchange genetic information, we call it <u>genetic recombination</u>, and it can occur in one of three ways: <u>conjugation</u>, <u>transformation</u>, or <u>transduction</u>.

15. In <u>conjugation</u>, bacteria link together to exchange circular strands of DNA called <u>plasmids</u>. In <u>transformation</u>, a bacterium can absorb a segment of <u>DNA</u> from a non-functional <u>bacterium</u>. In <u>transduction</u>, DNA can be transferred from one bacterium to another by a virus.

16. Bacteria can survive harsh conditions by forming <u>endospores</u>.

17. A <u>bacterial colony</u> is really just a simple association of individual bacteria. Bacteria in a streptococcus colony have a <u>spherical (coccus)</u> shape, while bacteria in a diplobacillus colony have a <u>rod (bacillus)</u> shape.

18. After a <u>Gram stain</u>, certain bacteria look blue when viewed under a microscope whereas others looked <u>red</u>.

19. <u>Gram-negative</u> bacteria belong in phylum Gracilicutes, while <u>Gram-positive</u> bacteria belong in phylum Firmicutes. Bacteria with cell walls significantly different from those in these two phyla belong in phylum <u>Mendosicutes</u>, while bacteria with no cell walls belong in phylum <u>Tenericutes</u>.

20. Phylum Gracilicutes has three classes: <u>Scotobacteria</u> (non-photosynthetic bacteria), <u>Anoxyphotobacteria</u> (photosynthetic bacteria that do not produce oxygen), and <u>Oxyphotobacteria</u> (photosynthetic bacteria that produce oxygen). Phylum Firmicutes has two classes: <u>Firmibacteria</u> (cocci and bacilli bacteria) and <u>Thallobacteria</u> (bacteria of any other shape). Phylum Tenericutes has only one class: <u>Mollicutes</u>. Phylum Mendosicutes has only one class: <u>Archaebacteria</u>. Many places that are uninhabitable to other organisms will be populated with members of class <u>Archaebacteria</u>.

21. Photosynthetic organisms called blue-green algae are more properly called <u>cyanobacteria</u>.

22. *Clostridium botulinum* can cause <u>botulism</u>. Undercooked eggs and poultry can give you *Salmonella* poisoning. *Escherichia coli* bacteria live in your gut. There are pathogenic <u>strains</u> and non-pathogenic <u>strains</u> of this bacterium.

23. For optimum growth, most bacteria need <u>Moisture</u>, <u>Moderate temperature</u>, <u>Nutrition</u>, <u>Darkness</u>, and <u>the proper amount of oxygen</u>.

Labeling exercise:

a. <u>plasma membrane</u> b. <u>flagellum</u> c. <u>capsule</u> d. <u>DNA</u> e. <u>cytoplasm</u> f. <u>cell wall</u> g. <u>fimbria</u>
h. <u>ribosome</u>

ANSWERS TO THE SUMMARY OF MODULE #3

1. Kingdom Protista is divided into two main groups: <u>protozoa</u> (mostly individual, single-celled creatures with a form of locomotion) and <u>algae</u> (mostly colonies of eukaryotic cells that have no form of locomotion).

2. Protozoa are split into four major phyla based on their locomotion: Mastigophora contains those that use <u>flagella</u>, Sarcodina contains those that use <u>pseudopods</u>, Ciliophora contains those that use <u>cilia</u>, and Sporozoa contains those that have <u>no means of locomotion</u>.

3. Algae are split into five major phyla based on habitat, organization, and cell wall. Chlorophyta contains those that live in <u>fresh water</u>, are composed of <u>single cells</u>, and have cell walls made of <u>cellulose</u>. Chrysophyta contains those that live in <u>fresh water and marine waters</u>, are composed of <u>single cells</u>, and have cell walls made of <u>silicon dioxide</u>. Pyrrophyta contains those that live in <u>marine waters</u>, are composed of <u>single cells</u>, and have cell walls made of <u>cellulose or something atypical</u>. Phaeophyta contains those that live in <u>cold marine waters</u>, are composed of <u>multiple cells</u>, and have cell walls made of <u>cellulose and alginic acid</u>. Rhodophyta contains those that live in <u>warm marine waters</u>, are composed of <u>multiple cells</u>, and have cell walls made of <u>cellulose</u>.

4. The main portion of a cell's DNA is stored in its <u>nucleus</u>. Membrane-bounded "sacs" in a cell are called <u>vacuoles</u>. Two main types of vacuoles are <u>food vacuoles</u>, which store food, and <u>contractile vacuoles</u>, which regulate the amount of water in the cell.

5. The cytoplasm in a cell can be split into <u>ectoplasm</u>, which is thin and watery, and <u>endoplasm</u>, which is more dense.

6. *Amoeba proteus* is a typical member of phylum <u>Sarcondina</u>, and it can form <u>cysts</u> to survive extreme conditions.

7. Genus *Euglena* contains organisms from phylum <u>Mastigophora</u>. When it comes to food, these creatures are both <u>heterotrophic</u> and <u>autotrophic</u>. They have firm but flexible shape-sustaining <u>pellicles</u> and a light-sensitive region known as an <u>eyespot</u>.

8. Photosynthesis requires a pigment called <u>chlorophyll</u>, which cells store in <u>chloroplasts</u>.

9. Round, green colonies found in phylum Mastigophora are found in genus *Volvox*.

10. When organisms form a relationship in which at least one of them benefits, it is called <u>symbiosis</u>. If all organisms involved benefit, it is specifically known as <u>mutualism</u>. If one benefits and the other neither benefits nor is harmed, it is specifically known as <u>commensalism</u>. If one benefits and the other is harmed, it is specifically known as <u>parasitism</u>.

11. Genus *Paramecium* contains organisms from phylum <u>Ciliophora</u>. Organisms in this genus have two <u>nuclei</u>. The <u>macronucleus</u> is the larger of the two, and it controls metabolism, while the <u>micronucleus</u> is the smaller of the two, and it controls reproduction.

12. Paramecia can exchange DNA through <u>conjugation</u>, but unlike this process in bacteria, the DNA exchange is <u>mutual</u>.

13. Genus *Plasmodium* contains organisms from phylum <u>Sporozoa</u> that cause <u>malaria</u>. The organisms are transferred between people by the action of <u>mosquitoes</u>.

14. Members of phylum Sporozoa form <u>spores</u> as a part of their normal life cycle.

15. Tiny organisms that float in the water are called <u>plankton</u>. Small animals and protozoa are called <u>zooplankton</u>, while photosynthetic organisms (typically algae) are called <u>phytoplankton</u>.

16. When conditions are ideal, algae will reproduce so rapidly that they essentially "take over" their habitat. This is referred to as an <u>algal bloom</u>.

17. Members of phylum Chlorophyta have the pigment <u>chlorophyll</u> and are often called <u>green algae</u>.

18. <u>Cellulose</u> is a compound made of certain types of sugars that is common in many cell walls.

19. The members of phylum Chrysophyta are often called <u>diatoms</u> and are responsible for a large amount of the photosynthesis that occurs in creation. When the cell wall remains of many of these organisms clump together, they form a crumbly, abrasive substance called <u>diatomaceous earth</u>.

20. A <u>sessile colony</u> is a colony that does not move and anchors itself to an object with a <u>holdfast</u>.

21. Members of phylum Pyrrophyta are often referred to as <u>dinoflagellates</u>. They have two <u>flagella</u>. One species in this phylum, *Gymnodinium brevis*, have blooms that are called <u>red tides</u>.

22. Members of phylum Phaeophyta are often referred to as <u>brown algae</u>. Their cell walls contain <u>alginic acid (algin)</u> that is used as a thickening agent.

23. Members of genus *Macrocytis* in phylum Phaeophyta are often called <u>kelp</u> or <u>seaweed</u>. They form <u>holdfasts</u> that allow them to anchor themselves to rocks which sit at the bottom of the ocean. Some can grow as long as 100 meters.

24. Members of phylum Rhodophyta are often called <u>red algae</u>.

ANSWERS TO THE SUMMARY OF MODULE #4

1. The five features common to most fungi are <u>saprophytic feeding</u>, <u>extracellular digestion</u>, <u>reproduction by spores</u>, <u>multicellular makeup (you could say hyphae instead)</u>, and <u>cell walls containing chitin.</u>

2. The cells in some hyphae are separated by cell walls. Thus are called <u>septate hypha</u>. Other hyphae, called <u>nonseptate hyphae</u>, have no separations between the cells, and the nuclei are spread throughout the hypha. Even septate hyphae have <u>pores</u> through which <u>cytoplasm</u> is exchanged.

3. There are many forms of specialized hyphae. <u>Rhizoid hyphae</u> are imbedded in the material on which the fungus grows. <u>Aerial hyphae</u> are not imbedded in the material on which the fungus grows. If such a hypha produces spores, it is a <u>sporophore,</u> but if it asexually produces more filaments, it is a <u>stolon.</u> In the case of a fungus that feeds on a living organism, a hypha that enters the cells of the living organism and draws nutrients directly from the cytoplasm of those cells is called a <u>haustorium.</u>

4. All fungi are assumed (but not all are confirmed) to have some <u>sexual</u> mode of spore formation.

5. Sexual reproduction in fungi usually involves forming specialized spore-forming structures called <u>fruiting bodies,</u> which are the result of sexual reproduction between compatible <u>hyphae</u>.

6. One mode of asexual reproduction in fungi involves the lengthening of a <u>stolon</u>. After it reaches a certain length, it will begin to reproduce into hyphae that will form the <u>mycelium</u> of a new fungus.

7. There are six major phyla in kingdom Fungi. Members of phylum Basidiomycota form sexual spores on clublike <u>basidia</u>. Members of phylum Ascomycota form sexual spores in saclike <u>asci</u>. Members of phylum Zygomycota form sexual spores <u>where hyphae fuse</u>. Members of phylum Chytridiomycota form spores with <u>flagella</u>. Members of phylum Deuteromycota have no <u>known form of sexual reproduction</u> and are sometimes called <u>imperfect fungi</u>. Members of phylum Myxomycota are placed in kingdom <u>Protista</u> by some biologists, because they resemble <u>protozoa</u> for much of their lives.

8. Mushrooms make up most of the organisms in phylum <u>Basidiomycota</u>. In the mushroom life cycle, mycelia grow from <u>spores</u>. Two mycelia will <u>fuse</u> for sexual reproduction. The resulting mycelium will grow, crowding out the <u>parent mycelia</u>. Eventually, many hyphae will enclose themselves in a membrane, forming the <u>button</u> stage of the mushroom. A <u>fruiting body</u> will eventually emerge from the membrane, and it will be composed of three basic components: <u>stipe</u>, <u>cap</u>, and <u>gills</u>. The spores are formed on basidia found in the <u>gills</u>. Those basidia release spores, from which new mycelia will grow.

9. The mycelium of a mushroom tends to grow outward in <u>circular</u> patches. Because it can run out of food in the center, the fruiting bodies of the fungus can sometimes form a <u>fairy ring</u>.

10. <u>Puffball fungi</u> are part of phylum Basidiomycota and grow spores inside a membrane. When disturbed, the spores are pushed out a <u>hole</u> near the top of the membrane. <u>Shelf fungi</u> are also in this phylum and grow shelflike structures on both dead and living wood.

11. Rusts and smuts are examples of <u>parasitic</u> fungi in phylum Basidiomycota. Such a fungus tends to live most of its life cycle on one host, which is called its <u>main host</u>. However, it must also spend a certain part of its life cycle on an <u>alternate host</u>.

12. <u>Yeast</u> are single-celled members of phylum Ascomycota that asexually reproduce by <u>budding</u>. They are used in <u>fermentation</u>, which anaerobically breaks down sugars into smaller molecules, such as <u>carbon dioxide</u> and <u>alcohol</u>. The former can make bread dough <u>rise</u>, and the latter can be used to make <u>alcoholic</u> beverages.

13. Other members of phylum Ascomycota include <u>morels</u> (that have fruiting bodies which look like sponges), <u>cup fungi</u> (that have fruiting bodies which look like cups), <u>ergot of rye or *Claviceps purpurea*</u> (that feeds on rye grain and is deadly to people), and tree parasites that cause <u>Dutch elm disease</u> and <u>chestnut blight</u>.

14. The spores formed by fungi in phylum Zygomycota are called <u>zygospores</u>, which are composed of a hard, protective coating around a <u>zygote</u>.

15. Bread <u>mold</u> is in phylum Zygomycota. It forms <u>asexual</u> spores in sporangiophores.

16. Fungi in genus *Penicillium* are in phylum Deuteromycota. They produce the first <u>antibiotic</u> ever discovered, which revolutionized the treatment of infections. Some bacteria have adapted to be <u>immune</u> to these infection-fighting compounds, which means new ones must continually be found.

17. Members of phylum Myxomycota are often called <u>slime molds</u>. They tend to resemble <u>protozoa</u> during their feeding stage, and during this time, the mass of living tissue is called a <u>plasmodium</u>. During their reproductive stage, these organisms resemble <u>fungi</u>.

18. There is no such thing as an <u>unbiased</u> scientist.

19. A <u>lichen</u> is a <u>mutualistic</u> relationship between a fungus and an alga. The alga produces <u>food</u> for itself and the fungus by means of <u>photosynthesis</u>, while the fungus gives <u>supports</u> and <u>protection</u> to the alga. Most lichens reproduce by releasing a dust-like substance called a <u>soredium</u>, which contains spores of *both* the <u>alga</u> and the <u>fungus</u> in a protective case.

20. The <u>mycorrhiza (or fungus root)</u> is a mutualistic relationship between a fungus and a plant root. The fungus absorbs <u>nutrients</u> from the roots and gives the plant <u>minerals</u> in return.

ANSWERS TO THE SUMMARY OF MODULE #5

1. <u>Atoms</u> are the basic building blocks of matter. They are composed of <u>protons</u> and <u>neutrons</u> that form the <u>nucleus</u> at the center of the atom as well as <u>electrons</u> that orbit the nucleus. Atoms have an equal number of <u>protons</u> and <u>electrons</u>, and the majority of an atom's properties are determined by the number of <u>electrons</u> it has.

2. An <u>element</u> is a collection of atoms that all have the same number of protons.

3. A carbon atom has six protons and eight neutrons. The complete name of this atom is <u>carbon-14</u>, and it has <u>six</u> electrons.

4. The more important biological elements and their abbreviations (in parentheses) are: carbon <u>(C)</u>, <u>hydrogen</u> (H), oxygen <u>(O)</u>, <u>nitrogen</u> (N), phosphorus <u>(P)</u>, and <u>sulfur</u> (S).

5. "Sulfur-32" is the name of a specific <u>atom</u>, while "sulfur" is the name of an <u>element</u>.

6. When atoms link together, they form <u>molecules</u>. A molecule of ethyl alcohol, C_2H_6O, has <u>two</u> carbon atoms, <u>six</u> hydrogen atoms, and <u>one</u> oxygen atom.

7. Even though they both contain <u>carbon</u> and <u>oxygen</u> atoms, CO and CO_2 are <u>different</u> molecules.

8. When sucrose is dissolved in water, a <u>physical</u> change has taken place. On the other hand, when sucrose reacts with water with the help of an enzyme to make glucose and fructose, a <u>chemical</u> change has occurred. In general, <u>physical changes</u> are reversible, while <u>chemical changes</u> are not.

9. All matter can exist in one of three phases: <u>solid</u>, <u>liquid</u>, and <u>gas</u>. Adding energy turns <u>solids</u> into <u>liquids</u> and <u>liquids</u> into <u>gases</u>, while taking away energy tends to reverse the process.

10. When salt is dissolved in water, <u>salt</u> is the solute, <u>water</u> is the solvent, and the solution is <u>salt water</u>.

11. When a solute travels across a membrane in order to even out concentration, <u>diffusion</u> has occurred. When the solvent travels across a membrane in order to even out concentration, <u>osmosis</u> has occurred. <u>Osmosis</u> happens when a semipermeable membrane separates two solutions.

12. A cell sits in a solution that has a higher concentration of solutes than that found in the cell. Water will tend to travel <u>out of the cell and into the solution</u>.

13. In the balanced chemical equation:

$$C_{18}H_{32}O_{16} + 2H_2O \rightarrow 3C_6H_{12}O_6$$

<u>One</u> molecule of $C_{18}H_{32}O_{16}$ reacts with <u>two</u> molecules of H_2O to make <u>three</u> molecules of $C_6H_{12}O_6$.

14. Photosynthesis requires <u>carbon dioxide</u>, <u>water</u>, <u>energy from light</u>, and <u>chlorophyll</u> (which acts as a catalyst). It produces <u>glucose</u> and <u>oxygen</u> via the chemical equation:

$$6CO_2 + 6H_2O \rightarrow C_6H_{12}O_6 + 6O_2$$

15. Of the following molecules: $NaNO_3$, CH_2O, $C_6H_{15}N$, and KSCN, <u>CH_2O</u> and <u>$C_6H_{15}N$</u> are organic.

16. Glucose and fructose both have the same chemical formula, <u>$C_6H_{12}O_6$</u>, which means they are <u>isomers</u>. They have different <u>structural</u> formulas. A molecule can have more than one <u>structural</u> formula. Glucose and fructose, for example, have both a <u>ring</u> structure and a <u>chain</u> structure.

17. A simple sugar is called a <u>monosaccharide</u>. Two such simple sugars can join to make a <u>disaccharide</u>. If three or more join, they form a <u>polysaccharide</u>. Simple sugars join together through <u>dehydration</u> reactions.

18. People and animals store excess sugars as a <u>polysaccharide</u> known as <u>glycogen</u>. When they need the simple sugars again, they break down this molecule into <u>monosaccharides</u> via <u>hydrolysis</u> reactions.

19. The pH scale runs from <u>zero</u> to <u>fourteen</u>. A pH of <u>seven</u> is neutral. A pH lower than <u>seven</u> indicates an <u>acidic</u> solution, while a pH greater than <u>seven</u> indicates an <u>alkaline</u> solution.

20. Lipids are formed in <u>dehydration</u> reactions where three <u>fatty acid molecules</u> are joined to one <u>glycerol</u> molecule. Lipids are <u>hydrophobic</u>, meaning they are not attracted to water. If the <u>fatty acid molecules</u> that make up the lipid have no double bonds between the carbon atoms, it is a <u>saturated fat</u> and is generally <u>solid</u> at room temperature. If there are double bonds between the carbon atoms, it is an <u>unsaturated fat</u> and is generally <u>liquid</u> at room temperature.

21. Proteins are formed in <u>dehydration</u> reactions where <u>amino acids</u> are joined together. The bond that forms between them is called a <u>peptide bond</u>. <u>Enzymes</u> make up a special class of proteins that serve as <u>catalysts</u> for many biologically-important chemical reactions, and they typically work according to the <u>lock and key theory of enzyme action</u>, in which an active site complements the shape of a reactant. Many of these molecules are quite <u>fragile</u>, breaking down soon after they are formed.

22. <u>DNA</u> is a double chain of chemical units known as <u>nucleotides</u> that twist around one another in a double helix. The units that make up these chains are composed of three basic constituents: <u>deoxyribose</u>, a <u>phosphate group</u>, and a <u>nucleotide base</u>. The double helix is held together by <u>hydrogen bonds</u> that link certain <u>nucleotide bases</u> together. In DNA, <u>guanine</u> can link only to <u>cytosine</u> (and vice-versa), while the nucleotide base <u>adenine</u> can link only to <u>thymine</u> (and vice-versa).

Identification:

a. Fat (It has three fatty acid molecules linked to a glycerol.)

b. Monosaccharide (note the chemical formula: $C_4H_8O_4$. It has C and twice as many H's as O's, which makes it a carbohydrate. It is a small, simple, molecule, and simple carbohydrates are monosaccharides.)

c. Acid (It has an acid group and does not have the configuration of a fat.)

ANSWERS TO THE SUMMARY OF MODULE #6

Matching

1. j	6. n	11. d	16. f
2. q	7. s	12. o	17. p
3. k	8. b	13. i	18. m
4. a	9. g	14. h	19. c
5. t	10. r	15. l	20. e

Fill in the blanks.

21. Cells must perform at least eleven main functions in order to support and maintain life: <u>absorption</u>, <u>digestion</u>, <u>respiration</u>, <u>biosynthesis</u>, <u>excretion</u>, <u>egestion</u>, <u>secretion</u>, <u>movement</u>, <u>irritability</u>, <u>homeostasis</u>, and <u>reproduction</u>.

22. When a cell is placed in an isotonic solution, water <u>diffuses back and forth across the membrane</u>. When a cell is placed in a hypertonic solution, water <u>leaves the cell and enters the surroundings</u>, which can result in <u>plasmolysis</u>. When a cell is placed in a hypotonic solution, water <u>enters the cell from the surroundings</u>, which can result in <u>cytolysis</u>.

23. Cells store energy in little "packets" by converting <u>ADP</u> and <u>phosphate</u> to <u>ATP</u>. A gentle release of energy occurs when the <u>ATP</u> is converted back to <u>ADP</u> and <u>phosphate</u>.

24. Aerobic cellular respiration occurs in four steps. The first is <u>glycolysis</u>, in which <u>glucose</u> is converted to <u>two pyruvic acid molecules</u> and <u>four hydrogen</u> atoms. This takes <u>two ATPs</u> of energy, but it produces <u>four ATPs</u> of energy, for a net gain of <u>two ATPs</u>. The next step, the <u>formation of acetyl coenzyme A</u>, reacts <u>two molecules of pyruvic acid</u> with <u>two molecules of coenzyme A</u> to make <u>two molecules of acetyl coenzyme A</u>, <u>two molecules of carbon dioxide</u>, and <u>two hydrogen</u> atoms. It produces <u>zero ATPs</u> of energy. The third step, <u>the Krebs cycle</u>, reacts <u>two molecules of acetyl coenzyme A</u> and <u>three molecules of oxygen</u> to make <u>four molecules of carbon dioxide</u>, <u>two molecules of coenzyme A</u>, and <u>six hydrogen</u> atoms. It produces <u>two ATPs</u> of energy. The final step, <u>the electron transport system</u>, takes the <u>twelve hydrogen</u> atoms made in the previous steps and reacts them with <u>three oxygen molecules</u> to make <u>six water molecules</u>. It produces <u>32 ATPs</u> of energy.

25. If an animal cell performs respiration in aerobic conditions, it can make <u>36 ATPs</u> for each glucose molecule. Under anaerobic conditions, it can make only <u>two ATPs</u> for each glucose molecule.

Identification

a. <u>phospholipid</u> (You could be very precise and say the hydrophilic end of a phospholipid.)
b. <u>protein</u> (You could say active transport site)   c. <u>glycoprotein</u>   d. <u>carbohydrate</u>   e. <u>cholesterol</u>
f. <u>filaments of the cytoskeleton</u> g. <u>glycolipid</u>

ANSWERS TO THE SUMMARY OF MODULE #7

1. In determining your traits, your <u>DNA</u> sets a range of possible characteristics, called your <u>genetic tendency</u>, and your activities determine what portion of that range is manifested in your body.

2. The three factors that determine the characteristics of a person are <u>genetic factors</u>, <u>environmental factors</u>, and <u>spiritual factors</u>.

3. Of the three factors listed above, <u>genetic factors</u> are laid down first. The information in an organism's DNA is split up into little groups known as <u>genes</u>.

4. By and large, the tasks that a cell can complete are dependent on the <u>proteins</u> that it produces.

5. Like DNA, the nucleotides of <u>RNA</u> join together in long strands. Unlike DNA, however, they do not form a <u>double helix</u>. Like DNA, <u>RNA</u> has four nucleotide bases: <u>adenine</u>, <u>cytosine</u>, <u>guanine</u>, and <u>uracil</u>. Of these four nucleotide bases, <u>adenine</u> links to <u>uracil</u> (and vice-versa), and <u>cytosine</u> links to <u>guanine</u> (and vice-versa). Unlike DNA, the sugar that makes up the foundation of the nucleotides in <u>RNA</u> is <u>ribose</u>.

6. One of RNA's jobs is to take a "negative image" of the cell's DNA out of the <u>nucleus</u> and to the <u>ribosome</u>. The RNA that does this job is called <u>messenger RNA (mRNA)</u>.

7. Protein synthesis in the cell can be split into two basic steps: <u>transcription</u> and <u>translation</u>.

8. In <u>transcription</u>, <u>mRNA</u> make a "negative image" of the cell's DNA. It does this by building a molecule that has <u>adenine</u> anywhere the DNA has a thymine, <u>uracil</u> anywhere DNA has an adenine, <u>cytosine</u> anywhere that DNA has a guanine, and <u>guanine</u> anywhere that DNA has a cytosine. The newly-built <u>mRNA</u> molecule then leaves the <u>nucleus</u> and heads to the <u>ribosome</u>.

9. The information in the <u>mRNA</u> that leaves the nucleus is contained in a three-nucleotide-base sequence called a <u>codon</u>. Each <u>codon</u> specifies an <u>amino acid</u> in the protein that is to be made.

10. Translation occurs at the <u>ribosome</u>, which is surrounded by a lot of RNA molecules that have an <u>amino acid</u> attached. This type of RNA is called <u>transfer RNA (tRNA)</u>. The <u>tRNA</u> has a three-nucleotide-base sequence called an <u>anticodon</u>. If one of these <u>anticodons</u> can link to a <u>codon</u> on the mRNA, it will do so, pulling the <u>amino acid</u> along with it. This causes <u>amino acids</u> to line up in the particular sequence specified by the mRNA, and the amino acids are then linked to form a <u>protein</u>.

11. A <u>codon</u> that has the sequence adenine, uracil, guanine will attract a tRNA molecule with an <u>anticodon</u> that has the sequence <u>uracil</u>, <u>adenine</u>, <u>cytosine</u>.

12. Eukaryotic genes contain sections called <u>exons</u> (instructions for making a protein) and <u>introns</u> (best described as "spacers"). Between <u>transcription</u> and <u>translation</u>, the mRNA must be processed. This processing removes the <u>introns</u> and splices the <u>exons</u> together.

13. The <u>DNA</u> of a eukaryotic cell is tightly bound together with a network of proteins. Certain proteins, called <u>histones</u>, act as spools, which wind up small stretches of <u>DNA</u>. The DNA wrapped around these <u>histones</u> form what could be described as "beads on a string," which we call

nucleosomes. Other proteins stabilize and support these spools, making a complex network of DNA coils and proteins. This network is called a chromosome.

14. Asexual reproduction in eukaryotic cells is called mitosis. The stages of this process, in order, are prophase, metaphase, anaphase, and telophase. When the cell is not reproducing, it is said to be in interphase. This process takes one diploid cell and produces two diploid cells.

15. In a diploid cell, chromosomes come in homologous pairs. A haploid cell has only one member from each pair.

16. The familiar "X" shape a chromosome takes occurs because the chromosome is duplicated, and the duplicate is attached. Nevertheless, this still counts as only one chromosome.

17. Genetically, sex is determined by sex chromosomes, which typically come in two forms: X and Y.

18. Meiosis is the process by which gametes are produced for the purpose of sexual reproduction. It starts with one diploid cell and produces four haploid cells. In males, all four of the products are viable and are called sperm. In females, only one of the products is viable, and it is called an egg.

19. Meiosis occurs in two broad steps, called meiosis I and meiosis II. In the first step, the chromosomes are all duplicated, and then the homologous pairs are separated, resulting in two haploid cells with duplicated chromosomes. In the second step, the duplicates are separated from the originals, resulting in a total of four haploid cells with unduplicated chromosomes.

20. An organism has diploid cells that contain 20 chromosomes. If one of these cells undergoes mitosis, two cells with 20 chromosomes each will be produced. If it undergoes meiosis, four cells with 10 chromosomes each will be produced.

21. When two haploid cells fuse in fertilization, the result is a diploid cell, which is called a zygote.

22. A virus is a chemical entity that is not truly alive, but can infect cells via the lytic pathway. Although it is not alive, it has either DNA or RNA as its genetic material, and that material is housed in a protective protein coat. These chemical entities cause a wide range of diseases in people and animals. One particular type of virus, the bacteriophage, even infects bacteria.

23. Your body has several means by which it can protect itself from them, including phagocytic cells that engulf pathogens and antibodies, which are specialized proteins that help to ward off pathogens. Once your body produces antibodies against a particular pathogen, it can remember how to make them in case you are infected again. This is the principle behind a vaccine, which is one of the most effective means of protecting yourself against certain viruses.

ANSWERS TO THE SUMMARY OF MODULE #8

1. Gregor Mendel's life story shows what can happen when a person has a true desire to <u>learn</u>. His sacrifices for his <u>education</u> allowed him to unlock one of the deep <u>mysteries</u> of God's creation. Mendel's story also shows that when you <u>fail</u>, you should not give up. Finally, his willingness to put all of that away in order to defend the <u>faith</u> against an attack from the <u>government</u> shows that Mendel had the proper set of priorities.

2. Since animal cells have <u>homologous</u> pairs of <u>chromosomes</u>, we know genes come in <u>pairs</u>, with one <u>gene</u> on each <u>homologous</u> chromosome. Each gene that makes up one of these pairs is called an <u>allele</u>.

3. The <u>genotype</u> of an organism is essentially a list of its alleles. The <u>phenotype</u> of an organism is the observable expression of those alleles.

4. Mendel's principles in updated terminology:

1. <u>The traits of an organism are determined by its genes.</u>
2. <u>Each organism has two alleles that make up the genotype for a given trait.</u>
3. <u>In sexual reproduction, each parent contributes ONLY ONE of its alleles to its offspring.</u>
4. <u>In each genotype, there is a dominant allele. If it exists in an organism, the phenotype is determined by that allele.</u>

5. A diagram that follows a particular phenotype through several generations is called a <u>pedigree</u>.

6. When you cross two individuals concentrating on only one trait, you are performing a <u>monohybrid cross</u>. A <u>dihybrid cross</u> still deals with two individuals, but it concentrates on two separate traits.

7. Some traits are sex-linked, which means the alleles that define those traits are found on the <u>sex chromosomes</u> rather than the <u>autosomes</u>.

8. Many traits are caused by the interaction of *several* genes. This is called <u>polygenetic inheritance</u>. Some traits are controlled by alleles that exhibit <u>incomplete dominance</u>, which means the alleles tend to "mix" rather than one dominating the other. In some cases, one set of alleles might affect how another set of alleles is expressed. This is called <u>epistasis</u>. Sometimes, a single gene can affect multiple observable traits, which is called <u>pleiotrophy</u>. Often, there are <u>multiple alleles</u> for a gene, such as the human gene for blood type, which has A, B, and O alleles. In blood type, A and B are dominant over O but not dominant over each other. This is an example of <u>codominance</u>.

9. When an antigen is introduced into the blood, the body's response is to produce an <u>antibody</u>.

10. The "+" and "-" in blood type refers to the <u>Rh factor</u>, which is controlled by a single gene with two alleles. The "<u>+</u>" allele is dominant, while the "<u>-</u>" allele is recessive.

11. There are at least five means by which genetic abnormalities occur. In <u>autosomal inheritance</u>, a genetic abnormality is passed through the autosomes. In <u>sex-linked inheritance</u>, a genetic abnormality is passed through the sex chromosomes. In <u>allele mutation</u>, one of the alleles of a gene is chemically changed. In <u>change in the chromosome structure</u>, a chromosome can gain or lose genes. In <u>change in the chromosome number</u>, a cell can wind up with too many or too few of a specific chromosome.

Answer the following questions.

12. The first pea plant is homozygous (meaning both alleles are the same) and is tall. Thus, it must be TT. The other is heterozygous, which means it must have one of each. Thus, it must be Tt. The resulting Punnett square is:

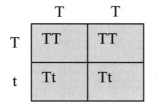

50% will have the genotype TT while 50% have the genotype Tt. 100% will have the phenotype of being tall.

13. The first pea plant is heterozygous, which means it must have one of each. Thus, it must be Yy. The other produces green peas. Since the green allele is recessive, the only way it can be expressed is if both of the plant's alleles are recessive. Thus, it is yy. The resulting Punnett square is:

	Y	y
y	Yy	yy
y	Yy	yy

This means 50% will be Yy and will make yellow peas, while 50% will be yy and will make green peas.

14. The first produces green peas. As mentioned above, that can only happen if it has both recessive alleles. In terms of texture, it is heterozygous, which means it has one of each. Thus, its genotype is yySs. The second is heterozygous in pea color, meaning it has one of each allele for color, and it expresses the recessive allele for wrinkled peas. Thus, it must have both recessive alleles for pea texture. As a result, its genotype is Yyss.

Now let's consider the possible gametes. In the first, the genotype is yySs. If the first "y" goes with the "S," the gamete will be yS. If the first "y" goes with the "s," the gamete will be ys. If the second "y" goes with the "S," the gamete will be yS. If the second "y" goes with the "s," the gamete will be ys. Thus, we have yS,ys,yS, and ys as the four possibilities from the first parent.

For the second parent, the genotype is Yyss. If the "Y" goes with the first "s," the gamete will be Ys. If the "Y" goes with the second "s," the gamete will be Ys. If the "y" goes with the first "s," the gamete will be ys. If the "y" goes with the second "s," the gamete will be ys. Thus, we have Ys,Ys,ys, and ys as the four possibilities from the second parent. This makes the Punnet square on the next page.

	yS	ys	yS	ys
Ys	YySs	Yyss	YySs	Yyss
Ys	YySs	Yyss	YySs	Yyss
ys	yySs	yyss	yySs	yyss
ys	yySs	yyss	yySs	yyss

The genotype YySs will produce yellow, smooth peas. Four of the 16 possibilities have this genotype, so 25% will produce yellow, smooth peas. The genotype Yyss will produce yellow, wrinkled peas. Four of the 16 possibilities have this genotype, so 25% will produce yellow, wrinkled peas. The genotype yySs will produce green, smooth peas. Four of the 16 possibilities have this genotype, so 25% will produce green, smooth peas. The genotype yyss produces green, wrinkled peas. Four of the 16 possibilities have this genotype, so 25% will produce green, wrinkled peas.

Please note that you could have solved this with a 2x2 Punnett square, because there are really only two possibilities for each gamete from each parent. If you don't understand this, don't worry. You don't need to. However, if you did solve it with a 2x2 Punnett square, your answers should be the same as those underlined above.

15. Since this is sex linked, the allele is on the X-chromosome, and the male has only one. Since he is red eyed, he must be $X^R Y$. The female is white eyed, so she must be $X^r X^r$, because the only way for a female to express the recessive allele is for both of her alleles to be recessive. The Punnett square, then, is:

	X^R	Y
X^r	$X^R X^r$	$X^r Y$
X^r	$X^R X^r$	$X^r Y$

Now remember, XY means male, and there are only two XYs. Both have the recessive allele, so all males will be white eyed. XX means female, and there are only two of them. Both have a dominant allele, so both are red eyed. Thus, 100% of the females will be red eyed, and 100% of the males will be white eyed.

16. Look at the parents on the left. They both can roll their tongues. However, one of the children cannot. What does this mean? It means that not rolling must be recessive. After all, to express the recessive trait, you must have 2 alleles. Thus, there must be one allele in each parent for the trait of not rolling the tongue. However, neither parent expresses that trait. If it were dominant, then the parents would have to express it if they each had one. However, they each have one but do not express it. That means the inability to roll the tongue is recessive.

We will say that the ability to roll the tongue, then, is R, and the inability is r.

<u>This means both of the first parents (1 and 2) must be Rr and Rr</u>. They each carry a recessive trait, but they do not express it. The male of the next (3) set is easy. He expresses the recessive trait. <u>Thus, he (3) must be rr</u>. Finally, the second parents produce offspring which also express the recessive trait. Thus, the mother must have one allele for it, or no offspring could express it. However, she does not express it, so the only possibility <u>for her (4) is Rr</u>.

17. Since the man is heterozygous in blood type, and is type A, his genotype must be AO. For the Rh factor, since he is heterozygous, he must be +-. Thus, he is AO+-. Since the woman is AB, her genotype for blood type is AB. Since she is Rh-negative, she expresses the recessive allele, meaning she must be --. Thus, she is AB--. This is a dihybrid cross (blood type and Rh factor), so let's look at the gametes.

For the man, the "A" can go with the "+", making A+. The "A" can also go with the "-", making A-. The "O" can go with the "+", making O+. The "O" can also go with the "-", making O-.

For the woman, the "A" can go with the first "-", making A-. The "A" can also go with the other "-", making A-.   The "B" can go with the first "-", making B-. The "B" can also go with the other "-", making B-.   The resulting Punnett square is:

	A+	A-	O+	O-
A-	AA+-	AA--	AO+-	AO--
A-	AA+-	AA--	AO+-	AO--
B-	AB+-	AB--	BO+-	BO--
B-	AB+-	AB--	BO+-	BO--

The genotype AA+- makes A+, since + is dominant. The genotype AO+- also makes A+, since A and + are dominant. That makes a total of four possibilities for A+. Thus, <u>25% of the children will have A+ blood</u>. The genotypes AA-- and AO-- make A- blood, since A is dominant over O. There are four possibilities there, so <u>25% of the children will be A-</u>. The genotype AB+- will make AB+ blood, since A and B are codominant. There are two possibilities, so <u>12.5% will be AB+</u>. The genotype AB-- makes AB- blood, so <u>12.5% will be AB-</u>. The genotype BO+- will make B+ blood, since B is dominant over O. There are two possibilities, so <u>12.5% will be B+</u>. The genotype BO-- makes B- blood, so <u>12.5% will be B-</u>.

ANSWERS TO THE SUMMARY OF MODULE #9

1. In 1859, Charles R. <u>Darwin</u> published a book entitled (in brief) *The Origin of Species*. In his book, he proposed a theory that attempted to explain the diversity of life on earth with no reference to God.

2. At one part in his life, Darwin believed the literal words of the <u>Bible</u>. However, he later compared it to the works of barbarians. The voyage of the <u>HMS Beagle</u> was a major turning point in his life.

3. Two scientists whose works had a profound effect on Darwin were <u>Thomas Malthus</u>, who studied populations, and <u>Charles Lyell</u>, who studied geology. <u>Malthus</u> inspired Darwin's idea that life is a constant <u>struggle for survival</u>, and <u>Lyell</u> impressed upon him that <u>the present is the key to the past</u>.

4. Darwin <u>did not</u> recant his theory and become a Christian on his deathbed.

5. Many science historians credit the <u>finches</u> that live on the <u>Galapagos archipelago</u> as inspiring Darwin's theory of evolution through natural selection.

6. According to Darwin, if an organism is born with a trait that makes it <u>more likely to survive</u>, it will most likely live long enough to reproduce, and it will pass that trait on to its offspring. This is called <u>natural selection</u>, because nature is "selecting" certain traits in certain species. Over time, this allows populations of species to <u>adapt</u> to changes in their surroundings.

7. Darwin masterfully showed that organisms do <u>adapt</u> to changes in their environments, which can cause some species to <u>change</u> significantly over time. This destroyed the age-old idea of the <u>immutability of species</u>, which essentially said that organisms never change significantly over time.

8. The idea that species can adapt to changes in the environment is referred to as <u>microevolution</u>, and it is a well-established scientific theory. The idea that one species can change into a radically different species over time is referred to as <u>macroevolution</u>, and it is (at best) an unconfirmed hypothesis.

9. Fossils are generally found in <u>sedimentary</u> rock, which is usually laid down in layers called <u>strata</u>.

10. The <u>geological column</u> is a hypothetical construct of what some geologists think all of the fossil-bearing strata in the world would look like if they <u>existed in one spot</u>. If you believe that rocks form according to the speculations of <u>Lyell</u>, you find that the geological column is evidence <u>for</u> evolution. If you believe that most of the fossil-bearing rocks that we see today were formed in global catastrophes such as a <u>worldwide flood</u>, the geological column is evidence <u>against</u> evolution.

11. The vast majority of the fossil record is made up of <u>clams and similar organisms</u>. What we see in a typical geological column represents only about <u>5%</u> of all fossils.

12. The myriad of transitional forms that the hypothesis of <u>macroevolution</u> predicts simply <u>cannot be found</u> in the fossil record. The few that <u>macroevolutionists</u> try to pass off as examples of transitional forms are so close to <u>one of the species they are trying to link</u> that their status of a transitional form is <u>hard to believe</u>. Two such examples are *Archaeopteryx*, which is supposed to link reptiles to birds, and *Australopithecus*, which is supposed to link apes to humans.

13. The fact that representatives from every major animal phylum can be found in Cambrian rock is called the <u>Cambrian explosion</u>. It is a real problem for macroevolution, because there is not nearly enough <u>time</u> in the Cambrian era to produce so much evolution, and there are no <u>transitional forms</u> linking one species to another.

14. The study of similar structures in different species is called <u>structural homology</u>, and it was once considered great evidence for evolution. After all, you could explain these similar structures by assuming that the species had a <u>common ancestor</u> which possessed the structure. However, it has been shown that similar structures are specified by <u>quite different</u> genes in different species. This tells us that they are not <u>inherited</u>. Thus, the study of similar structures in different species is now evidence <u>against</u> macroevolution.

15. There are certain <u>proteins,</u> such as hemoglobin and cytochrome C, that are found in most organisms. However, the sequence of <u>amino acids</u> in these chemicals is slightly different from one species to the next.

16. In comparing the cytochrome C found in bacteria, yeasts, fish, and horses, a macroevolutionist would expect that the <u>bacteria</u> and <u>yeasts</u> would have the most similar cytochrome C, and the <u>bacteria</u> and <u>horses</u> would have the most differences. In fact, <u>bacteria</u> and <u>horses</u> have cytochrome Cs that are slightly *more similar* than those of <u>bacteria</u> and <u>yeasts</u>. 99% of the data obtained by comparing similar proteins in difference species provide evidence <u>against</u> macroevolution.

17. Once the details of genetics were known, it was clear that the natural variations which occur between parent and offspring <u>could not</u> produce macroevolution. Thus, a new version of macroevolution, <u>neo-Darwinism</u> was formed. In this hypothesis, <u>mutations</u> caused radical changes between parent and offspring. Most such changes were <u>harmful</u> or <u>neutral,</u> but a few were <u>beneficial</u>. The <u>beneficial mutations</u> are supposed to have produced macroevolution. This hypothesis <u>reduced</u> the number of transitional forms expected in the fossil record, and it provided a means by which <u>information</u> could be added to the genetic code of a species.

18. Although it is possible for a <u>mutation</u> to make an organism more likely to survive under certain conditions, it does not <u>add information</u> to the genetic code. The mutations that have been observed to make an organism more likely to survive under certain conditions result in a <u>decrease</u> in information.

19. In the <u>punctuated equilibrium</u> hypothesis of macroevolution, mutations add genetic information to the genetic code, but they add it in steps that occur over <u>very short</u> time intervals. In between these time intervals, <u>no macroevolution</u> occurs. As a result, any transitional forms that result are <u>short-lived</u> and unlikely to be <u>fossilized</u>. This allows macroevolutionists to "explain away" the problem that there are <u>no real transitional forms</u> in the fossil record.

ANSWERS TO THE SUMMARY OF MODULE #10

1. Ecology is a study of the interactions between living and non-living things. In the taxonomy of ecological studies, a population is a group of interbreeding organisms coexisting together. A community is a group of populations living and interacting in the same area. An ecosystem is an association of living organisms and their physical environment. A biome is a group of ecosystems classified by climate and plant life.

2. The continent of Australia was nearly overrun by rabbits because a Rancher brought some from England to Australia for the purposes of hunting.

3. The four major trophic levels in a food web are producer (which makes its own food), primary consumer (which eats producers), secondary consumer (which eats primary consumers), and tertiary consumer (which eats secondary consumers). In general, an ecosystem has the greatest amount of biomass in producers and the least amount of biomass in tertiary consumers. This is because energy is lost as you travel up trophic levels in an ecosystem.

4. The bottom level of an ecological pyramid is always the widest level.

5. We studied three mutualistic symbiotic relationships in this module. The clownfish swims in the tentacles of the sea anemone. The former attracts prey to the latter, and the latter protects the former. There is also some evidence that the former protects the latter by scaring away one of its major predators. The goby lives in a hole that is dug by the blind shrimp. The former watches for predators while the latter cleans the hole. The blue-streak wrasse cleans the teeth of the Oriental sweetlips, and the material on the latter's teeth is food for the former. This kind of mutualism is very hard to explain in macroevolutionary terms.

6. If you were to try to explain the relationship between the blue-streak wrasse and the Oriental sweetlips using the hypothesis of macroevolution, you would be forced to believe that all of the instincts needed by the Oriental sweetlips and all of the instincts needed by the blue-streak wrasse evolved simultaneously in the two different animals.

7. The physical environment of an ecosystem is made up of all the non-living things in the ecosystem.

8. A watershed is an ecosystem where all water runoff drains into a single body of water.

9. In the water cycle, water can enter the atmosphere by evaporation and transpiration. It can leave the atmosphere by precipitation. In a watershed, the water can leave the ecosystem by river or stream flow. Because of the water cycle, water is continually transferred between the atmosphere and various bodies of water.

10. In the oxygen cycle, oxygen can enter the atmosphere by photosynthesis, water vapor destruction, and ozone destruction. It can leave the atmosphere through respiration, fire, rusting of metals, and ozone formation. The greatest contributor of oxygen to the atmosphere is photosynthesis, and the greatest removal of oxygen from the atmosphere comes as a result of respiration.

11. In the carbon cycle, carbon dioxide enters the atmosphere by <u>respiration</u>, <u>decomposition</u>, <u>fire</u>, and <u>fossil fuel burning</u>. It leaves the atmosphere by <u>dissolving into the ocean</u> and <u>photosynthesis</u>. <u>Respiration</u> puts the most carbon dioxide into the atmosphere, and <u>photosynthesis</u> takes the most out.

12. The <u>greenhouse effect</u> is the process by which energy that is being radiated by the earth is trapped in the atmosphere. This process is <u>essential</u> for the existence of life on earth.

13. Because people burn a lot of fossil fuels, the amount of <u>carbon dioxide</u> in the atmosphere is rising. This has caused some to worry about <u>global warming</u>.

14. Although <u>carbon dioxide</u> levels in the atmosphere have been rising steadily, the <u>average temperature</u> of the earth has not risen since 1925. A small increase (about 0.5 °C) occurred prior to 1925, but that was before <u>carbon dioxide</u> levels had risen significantly.

15. More than 200 scientific studies have shown that the average temperature of the earth was higher in the <u>Middle Ages</u> than it is today.

16. Although parts of the earth are getting warmer, other parts of the earth are getting <u>cooler</u>. As a result, the <u>average temperature</u> has not changed significantly over the past 75 years.

17. The process by which nitrogen gas (N_2) is converted to a form that is more useful to most organisms on the planet is called <u>nitrogen fixation</u>. Although some of this is done by the <u>physical environment</u>, most if it is done by <u>nitrogen-fixing bacteria</u>.

18. The more biologically-useful forms of nitrogen are <u>nitrates</u>, <u>nitrites</u>, and <u>ammonia</u>. These are produced by <u>lightning</u>, <u>nitrogen-fixing bacteria</u>, <u>decomposition</u>, and the <u>wastes</u> of living organisms. Sometimes, these more biologically-useful forms of nitrogen are converted back into nitrogen gas. This process is called <u>denitrification</u>.

19. <u>Producers</u> absorb the biologically-useful forms of nitrogen for their biosynthesis. Consumers get the nitrogen they need by <u>eating producers</u>.

ANSWERS TO THE SUMMARY OF MODULE #11

1. Although not official taxonomy groups, biologists use the terms <u>vertebrates</u> and <u>invertebrates</u> to refer to animals with and without backbones, respectively.

2. The three basic forms of symmetry among organisms in creation are <u>spherical</u> (any cut through the center produces two equal halves), <u>radial</u> (and longitudinal cut through the center produces two equal halves), and <u>bilateral</u> (only one longitudinal cut through the center produces two equal halves). A round organism like *Volvox* has <u>spherical symmetry</u>, a sea anemone has <u>radial symmetry</u>, and people have <u>bilateral symmetry</u>.

3. Phylum <u>Porifera</u> contains the sponges, which have <u>no</u> symmetry. These creatures pull <u>water</u> into their bodies, cleaning the water of <u>organisms and organic debris</u> that they eat. They are composed of an outer layer of cells called the <u>epidermis</u> and an inner layer of cells. These layers are separated by a jellylike substance called the <u>mesenchyme</u>. They support themselves with either <u>spicules</u> or <u>spongin</u>, the latter of which results in <u>soft</u> sponges. Water is pulled through their bodies by <u>collar cells</u>, and specialized cells called <u>amoebocytes</u> digest food and transport it throughout the body. In addition, they take in <u>waste products</u> from the inner cells and travel to the epidermis to release them. These cells also exchange <u>gases</u> with the surroundings and produce the lime or silica that makes up the <u>spicules</u>. Sponges can produce a <u>gemmule</u>, which can survive through a long period of inclement weather.

4. Phylum <u>Cnidaria</u> contains jellyfish, sea anemones, and hydra. These creatures have two basic forms: the <u>polyp</u> and the <u>medusa</u>. Some, like the jellyfish, take on <u>both</u> forms during different parts of their life cycles. The bodies of these creatures have two layers of <u>epithelium</u> separated by a jellylike layer called a <u>mesoglea</u>. They have <u>radial</u> symmetry, and their tentacles (and sometimes bodies) are covered with stinging <u>nematocysts</u>.

5. Hydra have the <u>polyp</u> form. They asexually reproduced by <u>budding</u>, and they sexually reproduce using <u>ovaries</u> that produce eggs and <u>testes</u> that produce sperm. When hydras were first discovered, they were thought to be a transitional form between <u>plants</u> and <u>animals</u>. We now know they are 100% <u>animals</u>.

6. While the <u>nematocysts</u> of the hydra are triggered by pressure, the <u>nematocysts</u> of the sea anemone are triggered by a complex <u>chemical recognition system</u>.

7. Corals are also members of phylum <u>Cnidaria</u> that use the <u>polyp</u> form.

8. In the life cycle of a jellyfish, sexual reproduction occurs in the <u>medusa</u> form, and asexual reproduction occurs in the <u>polyp</u> form.

9. Segmented worms are in phylum <u>Annelida</u>.

10. In order to move, the earthworm uses its posterior <u>setae</u> to anchor its posterior end while it contracts its <u>circular</u> muscles, stretching the earthworm out. This pushes the <u>anterior</u> end forward. Once these muscles have contracted completely, the earthworm then uses it anterior <u>setae</u> to anchor its anterior end releases its posterior <u>setae</u>. After that, the <u>longitudinal</u> muscles contract, causing the <u>posterior</u> end to move forward.

11. The earthworm ingests soil into its mouth with its <u>pharynx</u>. The soil is then passed into the <u>crop</u>, where it is stored for a while. Eventually, the soil makes it to the <u>gizzard</u>, where it is ground into small pieces. The worm <u>digests</u> edible materials in the <u>intestine</u>, and the inedible materials are pushed out through the <u>anus</u>. The digested food is absorbed by <u>blood</u> that circulates through the walls of the intestine. The blood <u>transports</u> the digested food to cells throughout the worm's body, pushed by contractions of the <u>aortic arches</u>. Some wastes are also gathered in small organs called <u>nephridia</u> which release the waste through tiny holes called <u>nephridiopores</u>.

12. Earthworms absorb oxygen and release carbon dioxide through a moist layer in the epidermis called a <u>cuticle</u>. Its nervous system is controlled by two <u>ganglia</u> in the anterior end of the body. Signals come to and from the ganglia through a <u>ventral nerve cord</u>.

13. Earthworms are <u>hermaphroditic</u>, meaning they have both male and the female reproductive organs. During mating, two earthworms deposit sperm (stored in their <u>seminal vesicles</u>) into each other's <u>seminal receptacles</u>. The sperm are later used to fertilize the eggs that are stored in the <u>oviducts</u>. The zygote produced develops in a <u>cocoon</u>.

14. Flatworms like the planarian belong in phylum <u>Platyhelminthes</u>. They reproduce asexually by ripping themselves in half and <u>regenerating</u>. They also <u>sexually</u> reproduce much like earthworms. Its complex nervous system allows the planarian to have senses of <u>taste</u>, <u>smell</u>, <u>touch</u>, and the ability to sense <u>light</u>. It senses <u>light</u> using two <u>eyespots</u> on its anterior end.

15. Phylum <u>Nematoda</u> contains roundworms. Their bodies are essentially composed of a <u>tube</u> within a <u>tube</u>, and many members of this phylum are <u>parasitic</u>.

16. Phylum <u>Mollusca</u> contains clams, oysters, snails, and squid. Most members of this phylum have a <u>mantle</u>, a <u>shell</u>, a <u>visceral hump</u>, a <u>foot</u>, and a <u>radula</u>. A mollusk with one shell is called a <u>univalve (gastropod)</u>, while one with two shells is called a <u>bivalve (pelecypod)</u>.

Identification

a. <u>mouth</u> b. <u>ventral nerve cord</u> c. <u>seminal receptacles</u> d. <u>seminal vesicles</u> e. <u>ventral blood vessel</u>
f. <u>nephridia with nephridiopores</u> g. <u>clitellum</u> h. <u>intestine</u> i. <u>dorsal blood vessel</u> j. <u>gizzard</u> k. <u>crop</u>
l. <u>oviduct</u> m. <u>ovary</u> n. <u>esophagus</u> o. <u>aortic arches</u> p. <u>pharynx</u> q. <u>ganglia</u>

ANSWERS TO THE SUMMARY OF MODULE #12

1. Arthropods have the following five characteristics in common: <u>an exoskeleton</u>, <u>body segmentation</u>, <u>jointed appendages</u>, <u>ventral nervous system</u>, and an <u>open circulatory system</u>.

2. Because they have an exoskeleton, arthropods must <u>molt</u> in order to grow. An arthropod's body might consist of three segments: <u>head</u>, <u>thorax</u>, and <u>abdomen</u>, or it might consist of two segments: <u>cephalothorax</u> and <u>abdomen</u>. The eyes might be <u>simple</u> (consisting of one lens) or <u>compound</u> (consisting of many lenses).

3. The crayfish belongs in class <u>Crustacea</u> of phylum Arthropoda. It gets its sense of balance mostly from its <u>antennules</u> and its sensitive senses of taste and touch from its <u>antennae</u>. It uses its <u>chelipeds</u> (claws) to grab onto prey and for defense. Its <u>uropod</u> and <u>telson</u> on its posterior end are used for swimming, and its <u>swimmerets</u> are used for swimming and reproduction.

4. A crayfish gets oxygen that has been dissolved in the water by passing water over its <u>gills</u>. Blood distributes the oxygen throughout the body. It collects in the <u>pericardial sinus</u> around the <u>heart</u>. It enters the <u>heart</u> through one of three openings which close when the <u>heart</u> is ready to pump. The <u>heart</u> pumps blood through blood vessels that are open at the other end. These vessels dump the blood <u>directly into various body cavities</u>. Gravity causes the blood to fall into the <u>sternal sinus</u>, where it is collected by blood vessels that are open at one end. These vessels carry the blood back towards the <u>pericardial sinus</u>. On its way there, the blood is passed through the <u>gills</u> where it can release <u>carbon dioxide</u> and absorb <u>oxygen</u>. The blood also passes through a <u>green gland</u>, which cleans it.

5. In order to eat, the crayfish uses its <u>mandibles</u> to break the food into small chunks. The food enters a short <u>esophagus</u> and goes into a <u>stomach</u> that has essentially two regions. The first region <u>grinds the food up into fine particles</u>. The second region <u>sorts</u> the particles. If they are small enough, they are sent directly to <u>digestive glands</u> which secrete enzymes, completing the digestion process. If the particles are too large to be digested immediately, they are sent to the <u>intestine</u>. Anything that remains at the other end of the intestine is considered indigestible and is expelled out the <u>anus</u>.

6. A crayfish's brain is comprised of two <u>ganglia</u> that each has a <u>nerve cord</u>.   At the base of each antennule is a <u>statocyst</u>, which gives the crayfish its sense of balance.

7. In male crayfish, <u>sperm</u> are formed in the <u>testes</u> and transferred to the first and second pairs of <u>swimmerets</u>. Male crayfish deposit their <u>sperm</u> into special containers that the female has. The female will then store the sperm until spring. In the spring, the female produces <u>eggs</u>. They are fertilized by the <u>sperm</u> and go to the <u>swimmerets</u>. The fertilized eggs <u>attach themselves to the swimmerets</u> and develop for approximately six weeks, at which point the eggs hatch.

8. Spiders are in class <u>Arachnida</u>. Members of this class have these five characteristics in common: <u>Four pairs of walking legs</u>, <u>a cephalothorax</u>, <u>four pairs of simple eyes</u>, <u>no antennae</u>, and <u>respiration done through book lungs</u>. They spin webs make of <u>silk</u> that is produced in <u>silk glands</u> and spun with <u>spinnerets</u>.

9. Centipedes, aggressive predators with <u>two</u> legs per segment, are found in class <u>Chilopoda</u>. Millipedes, docile herbivores with <u>four</u> legs per segment, are found in class <u>Diplopoda</u>.

10. Members of class Insecta have four characteristics in common: <u>three pairs of walking legs</u>, <u>wings at some stage of their life</u>, <u>one pair of antennae</u>, <u>three body segments</u>. They do not have lungs, but breathe through <u>tracheas</u> that form an intricate network throughout the body. They all go through some kind of <u>metamorphosis</u> in their life cycle. <u>Complete metamorphosis</u> has four stages: <u>egg</u>, <u>larva</u>, <u>pupa</u>, and <u>adult</u>. <u>Incomplete metamorphosis</u> has three stages: <u>egg</u>, <u>nymph</u>, and <u>adult</u>.

11. Butterflies and moths belong to order <u>Lepidoptera</u>, while the ants, bees, and wasps belong in order <u>Hymenoptera</u>. These are also called <u>social insects</u>, because they exist in a complex society. Beetles belong to order <u>Coleoptera</u>. Flies, gnats, and mosquitoes belong in order <u>Diptera</u>. Grasshoppers and crickets go in order <u>Orthoptera</u>.

Identification

Exterior view:

a. <u>antennae</u> b. <u>antennules</u> c. <u>cephalothorax</u> d. <u>abdomen</u> e. <u>telson</u> f. <u>uropods</u> g. <u>swimmerets</u>
h. <u>carapace</u> i. <u>walking legs</u> j. <u>chelipeds</u>

Interior view:

a. <u>eye</u> b. <u>brain ganglia</u> c. <u>stomach</u> d. <u>gonad</u> e. <u>heart</u> f. <u>pericardial sinus</u> g. <u>intestine</u> h. <u>anus</u>
i. <u>nerve cord</u> j. <u>digestive glands</u> k. <u>sternal sinus</u> l. <u>mouth</u> m. <u>esophagus</u> n. <u>green gland</u>

ANSWERS TO THE SUMMARY OF MODULE #13

1. Members of phylum Chordata have either a <u>backbone</u> or a <u>notochord</u>. The sea squirt is a member of subphylum <u>Urochordata</u>, and it has a <u>notochord</u> only in its larval stage. As an adult, it is supported by a leathery <u>tunic</u>. The lancelet is a member of subphylum <u>Cephalochordata</u>, and it has a <u>notochord</u> throughout its life. Animals with backbones are members of subphylum <u>Vertebrata</u>.

2. There are three types of bone cells: <u>osteoblasts</u>, <u>osteocytes</u>, and <u>osteoclasts</u>. The bones they comprise are made of fibers of a protein called <u>collagen</u> that have been hardened by <u>calcium-containing salts</u>. Bone tissue comes in two types: <u>compact bone tissue</u> (composed of tightly-packed fibers) and <u>spongy bone tissue</u> (composed of loosely-packed fibers). These tissues are surrounded by a membrane called the <u>periosteum</u> that contains blood vessels and nerves. In the very center of the bone is a cavity that holds <u>bone marrow</u>, which produces blood cells.

3. A typical vertebrate endoskeleton is made up of an <u>axial skeleton</u> (that supports and protects the head, neck, and trunk), and an <u>appendicular skeleton</u> (has the limbs attached to it).

4. The vertebrate circulatory system is <u>closed</u>, and it is composed of <u>veins</u> (that carry blood towards the heart), <u>arteries</u> (that carry blood away from the heart), and <u>capillaries</u> (that allow gases to be exchanged with the tissues). Oxygen is carried by <u>red blood cells</u> that get their color from <u>hemoglobin</u>.

5. Most vertebrate brains have five sets of lobes: <u>olfactory lobes</u> (for smell), <u>cerebrum lobes</u> (for integrating sensory information and making responses), <u>optic lobes</u> (for sight), <u>cerebellum</u> (for involuntary actions and refining muscle movement), and the <u>medulla oblongata</u> (for vital functions). Signals are sent down the medulla oblongata and into the <u>spinal cord</u>.

6. Vertebrate reproduction can occur by <u>internal</u> or <u>external</u> fertilization, and the embryo's development can be <u>oviparous</u> (in an egg outside the female's body), <u>ovoviviparous</u> (in an egg inside the female's body), or <u>viviparous</u> (inside the female, connected to her by a placenta).

7. Lamprey eels are in class <u>Agnatha</u>, members of which are commonly called the <u>jawless fish</u>. They are <u>anadromous</u>, which means they hatch in <u>fresh water</u>, migrate to <u>salt water</u> as adults, then go back to <u>fresh water</u> in order to reproduce.

8. Sharks, rays, and skates belong to class <u>Chondrichthyes</u>. They have <u>cartilaginous</u> endoskeletons, which are more flexible than human endoskeletons. Sharks hunt using many senses, including a <u>lateral line</u> that detects vibrations, and a keen <u>electrical field sensor</u> that detects minute amounts of electricity.

9. Members of class <u>Osteichthyes</u> are typically called "<u>bony fish</u>," since most or all of their skeletons are hardened with calcium. A member of this class gets its sense of taste from <u>taste buds</u> on its tongue. Food passes through a <u>pharynx</u> (throat) and into the <u>esophagus</u>, which leads to the <u>stomach</u>. There, the food is broken down and stored. It is then sent to the <u>intestine</u>, where it is digested. Any undigested remains leave the fish through the <u>anus</u>. The fish has a <u>liver</u> that produces <u>bile</u>, which is concentrated in the <u>gall bladder</u> and helps digest fats. The <u>air bladder</u> helps it control its depth in the water.

10. Bony fish have a <u>two-chambered</u> heart composed of an <u>atrium</u> and a <u>ventricle</u>. Blood from the tissues enters the <u>atrium</u> and is then transferred to the <u>ventricle</u>, which pushes it out. It then passes

over the gills to get oxygen and release carbon dioxide. It then goes to the tissues. While it passes through the body, it gets cleaned in organs called kidneys.

11. All of the organisms in this module are ectothermic, which means they are "cold-blooded."

12. Amphibians have these six characteristics in common: endoskeleton made mostly of bone, smooth skin with many capillaries and pigments (no scales), two pairs of limbs with webbed feet, as many as four organs for respiration, three-chambered heart, and oviparous with external fertilization.

13. In its larval stage, an amphibian breathes with gills. As an adult, it breathes with lungs, its skin, and the lining of its mouth. It deals with cold weather by hibernating.

Identification

Organs:

a. esophagus b. brain c. spinal cord d. stomach e. air bladder f. kidney g. gonad h. anus
i. intestine j. pyloric ceca k. gall bladder l. liver m. heart n. gills

Circulatory system:

a. anterior cardial vein b. efferent brachial arteries c. dorsal aorta d. kidney
e. posterior cardial vein f. atrium g. ventricle h. ventral aorta i. afferent brachial arteries j. gills

ANSWERS TO THE SUMMARY OF MODULE #14

1. The study of plants is called <u>botany</u>. Plants that grow year after year are <u>perennials</u>, while plants that live for only one year are called <u>annuals</u>. Plants that live for two years are <u>biennials</u>.

2. Plants have two types of organs: <u>vegetative</u> and <u>reproductive</u>. The former are considered vegetables, while the later are considered <u>fruits</u>. These organs are made of up to four different types of tissues: <u>meristematic</u> (containing undifferentiated cells), <u>ground</u> (the most common), <u>dermal</u> (the outer layer), and <u>vascular</u> (which carry water and nutrients). If the vascular tissue carries water and minerals, it is called <u>xylem</u> and is made of dead cells. If it carries food and other organic chemicals, it is called <u>phloem</u> and is made of living cells.

3. The primary portion of a leaf is called the <u>blade</u>, and the very tip of the blade is called the <u>apex</u>. The blade is attached to the stem with a small stalk called the <u>petiole</u>. At the base of the petiole, most plants have <u>stipules</u>. A <u>simple</u> leaf is one leaf attached to the stem of the plant by a single petiole. A <u>compound</u> leaf has several leaflets attached to a single petiole.

4. Plants in class <u>Monocotyledonae</u> (typically called <u>monocots</u>) have leaves with <u>parallel</u> venation, seeds with one cotyledon, typically have <u>fibrous root systems</u>, and produce flowers with petals in multiples of <u>three</u> or <u>six</u>. On the other hand, if the venation is <u>netted</u>, the plants belong to class <u>Dicotyledonae</u> and are typically called <u>dicots</u>. These plants have seeds with <u>two</u> cotyledons, have <u>taproot systems</u>, and produce flowers with petals that are usually in groups of <u>four</u> or <u>five</u>.

5. The top and bottom of a leaf are covered with a single layer of cells called the <u>epidermis</u>, which protects the inner parts of the leaf. Sometimes, the epidermis secretes a waxy substance called a <u>cuticle</u>. Tiny holes called <u>stomata</u> are found on the <u>underside</u> of most leaves. They allow for the exchange of <u>gases</u> with the atmosphere. Each stoma is flanked by two cells called <u>guard cells</u>, which open and close the stoma. Under the epidermis on both sides of the leaf are <u>parenchyma tissues</u>, which are composed of cells that do the photosynthesis. These tissues are composed of two layers, the <u>palisade mesophyll</u> (which has densely-packed cells) and the <u>spongy mesophyll</u> (which has loosely-packed cells). The veins in a leaf are made up of three tissues: <u>xylem</u>, <u>phloem</u>, and <u>collenchyma</u>.

6. Leaves have <u>chlorophyll</u>, which makes most of them green. Some leaves, however, have <u>carotenoids</u> as well, which have yellow or orange hues. In most leaves, the color of the <u>chlorophyll</u> overwhelms the colors of the <u>carotenoids</u>, but in some plants, the leaf picks up some color from them. In certain leaf tissues, there is another set of pigments called <u>anthocyanins</u>, which have different colors, depending on the pH of the leaf tissue. When a leaf falls off the tree and dies, the <u>chlorophyll</u> decays, allowing its other pigments to show. Plants that lose their leaves for the winter are called <u>deciduous</u>. At the base of each petiole in a deciduous tree, there is a thin layer of tissue called the <u>abscission layer</u>. As the days get shorter, this tissue blocks off the <u>xylem</u> and <u>phloem</u> running through the petiole.

7. Roots have three primary functions: <u>absorb water and nutrients</u>, <u>anchor the plant</u>, and <u>store food</u>.

8. There are basically two kinds of root systems: <u>fibrous</u> (which looks much like an underground bush) and <u>taproot</u> (in which the primary root continues to grow and stays the main root).

9. Longitudinally, a root is split into four regions: the root cap (made of dead, thick-walled cells), the meristematic region (where undifferentiated cells carry on mitosis), the elongation region (where cells differentiate and stretch), and the maturation region (where the cells are becoming fully differentiated). Root hairs, which increase the surface area of the root, are produced in the maturation region.

10. In a lateral cross section of a root, the epidermis is the outer layer. The cells inside the epidermis are called the cortex, where substances are stored for later use. Inside the root, there is another one-cell-thick layer called the endodermis. These cells surround the vascular chamber, which contains the xylem and phloem.

11. Plant stems can be either woody or herbaceous. They perform three basic functions: support and manufacture leaves, conduct water and nutrients to and from the leaves, carry on photosynthesis.

12. In herbaceous stems, dicots and monocots can be easily distinguished by looking at their fibrovascular bundles. The bundles of a monocot look like a face, and the bundles are distributed throughout the stem. The bundles of a dicot are found in a ring near the edge of the stem. In dicots, a vascular cambium can form new xylem or phloem, while a mature moncot has no such tissue.

13. Woody stems have an outer layer of bark that is actually composed of two layers: the inner bark (composed of phloem and cortex tissue) and the outer bark (composed of dead cork cells). Between these two layers is the cork cambium, that continually produces cork cells. The formation of bark allows a woody stem to continue to grow, unlike most herbaceous stems. New xylem and phloem must always be produced inside a woody stem. This is done by the vascular cambium, which produces phloem on its outer side and xylem on its inner side. This forms a pattern of alternating areas of light and dark wood, which forms what we call annual growth rings, which you can count to determine the age of the stem.

14. Some stems appear to be roots. Tubers are underground stems that store food, and bulbs are underground leaves that sprout from an underground stem.

15. Plants can be split into two basic groups: bryophytes (plants without vascular tissue) or tracheophytes (plants with vascular tissue). Bryophytes must be small.

16. Mosses are found in phylum Bryophyta and are composed of leafy shoots and rhizoids. They have an alternation of generations life cycle in which the plant we recognize as moss is the gametophyte generation and is composed of haploid cells. This generation produces gametes through mitosis. When fertilization occurs, the result is the sporophyte generation, which is made of diploid cells. This generation produces spores by meiosis, which germinate into the gametophyte generation. The gametophyte generation is dominant in mosses.

17. Ferns are members of phylum Pterophyta. They also have an alternation of generations life cycle, but the dominant generation is the sporophyte generation, which is made of diploid cells.

18. Evergreens are members of phylum Coniferophyta and produce seeds that are the result of fertilization between pollen made in pollen cones and eggs made in seed cones. Phylum Anthophyta contains the flower-making plants, which are either monocots or dicots.

ANSWERS TO THE SUMMARY OF MODULE #15

1. The study of life processes in an organism is called <u>physiology</u>.

2. In plants, water is used for essentially four processes: <u>photosynthesis</u>, <u>turgor pressure</u>, <u>hydrolysis</u>, and <u>transport</u>. One of these processes (<u>turgor pressure</u>) is responsible for a particular kind of motion in plants. This motion, which is in response to a stimulus but independent of the direction of the stimulus, is called <u>nastic movement</u>. The other type of motion, called a <u>tropism</u>, is not controlled by water and is dependent on the direction of the stimulus.

3. Plants store excess food in the form of a polysaccharide known as <u>starch</u>.

4. A good soil for plant growth is a <u>loam</u>, which is a mixture of gravel, sand, silt, clay, and organic matter. It should have large enough <u>pore spaces</u> to allow for plenty of oxygen, but small enough <u>pore spaces</u> to allow for the retention of water.

5. The <u>cohesion-tension theory</u> describes how water moves up a plant through the xylem. In this theory, <u>transpiration</u> causes a <u>tension</u>, pulling nearby water molecules up to fill the space left behind. The <u>cohesion</u> of the water molecules further down the xylem causes them to be pulled up as well.

6. The transport of organic chemicals and food through the phloem is called <u>translocation</u>, and it depends on the fact that the phloem are composed of <u>living</u> cells.

7. Controlling mitosis and regulating plant development is done by at least five groups of <u>hormones</u>. <u>Auxins</u> affect the way that cells elongate, which influences how the plant grows. <u>Phototropism</u> (Growing towards light), <u>gravitropism</u> (growing against gravity), and <u>thigmotropism</u> (a growth response to touch) are all controlled by these hormones. <u>Gibberellins</u> promote elongation in stems, affect mitosis rates, and induce seeds to germinate. <u>Cytokinins</u> affect mitosis rates, influence cellular differentiation, induce leaf cells to elongate, and affect the synthesis of chlorophyll. <u>Abscisic acid</u> inhibits the abscission layer so that it doesn't close off, and it also helps to control the stomata. <u>Ethylene</u> promotes the ripening of fruits and causes the abscission layer to close. In addition, botanists suspect there is at least one more <u>hormone</u>, called <u>florigen</u>, which controls flowering in an anthophyte.

8. <u>Insectivorous plants</u> trap and digest insects, but they <u>do not</u> use them for food. Instead, they use them to get materials for <u>biosynthesis</u> that cannot be found in the soil in which they grow.

9. Asexual reproduction in plants is typically called <u>vegetative reproduction</u>. Stems and roots can sometimes develop <u>roots</u> and grow into new plants if put in soil. Some plants produce <u>stems</u> that originate in their roots and become new plants. Other plants have specialized stems like <u>tubers</u> that can grow roots and become a new plant. Some plants produce <u>runners</u> that grow along the ground and then sprout a new plant on the end. Also, the stem of one plant can be <u>grafted</u> onto a different plant. When this happens, the stem is called the <u>scion</u> and the other plant is called the <u>stock</u>.

10. Phylum <u>Anthophyta</u> contains the flowering plants, and the flowers are the <u>reproductive organs</u>.

11. The female part of the flower is called the <u>carpel</u>, which is sometimes called the <u>pistil</u>. It is made up of three parts: <u>stigma</u>, <u>style</u>, and <u>ovary</u>. The male part of the flower is the <u>stamen</u>, and it is made up

of the anther and filament. If a flower has both, it is called a perfect flower. If it has one or the other, it is an imperfect flower.

12. The anther contains pollen sacs which produce pollen grains. Diploid cells in the pollen sacs go through meiosis to produce four haploid cells. Those cells each go through mitosis to produce a total of eight haploid cells. The cells are encased in four shells, and they differentiate into a tube nucleus and a sperm cell. In some plants, the sperm cell undergoes mitosis so that there are two sperm cells.

13. In the ovule, which is inside the ovary, a diploid cell undergoes meiosis to produce four haploid cells, three of which are polar bodies and die. The remaining cell, called the megaspore, undergoes mitosis without cell division three times to make eight haploid nuclei. Cell division then occurs, producing six small haploid cells and one large cell with two haploid nuclei.

14. To reproduce, a flower releases its pollen grains. They travel to another flower by wind, insects, or birds and land on the stigma of the carpel. The tube nucleus then digs a pollen tube through the style to the ovary. If there was only one sperm cell in the pollen grain, it undergoes mitosis to make two. Both sperm cells enter the ovule. One fertilizes one of the six small haploid cells, and the other fertilizes the large cell with two haploid nuclei. The former produces the zygote, and the latter produces the endosperm, which becomes the nutrient source for the developing plant embryo. This process, called double fertilization, is unique to kingdom Plantae.

15. When a seed is mature, it has either one or two cotyledons, which makes the plant a monocot or dicot. In some seeds, the cotyledon consumes the endosperm and becomes the food source for the embryo. In other seeds, it simply transfers nutrients from the endosperm to the embryo. The ovary then ripens into a fruit that encases the seed.

16. When the seed germinates, the radicle is the first to emerge, and it becomes the roots of the plant. The hypocotyl develops into the stem, and the first true leaves of the plant develop from the epicotyl. In some plants, the first leaves are not true leaves, but are "seed leaves," which are the cotyledons. Some cotyledons actually perform photosynthesis for a while until the true leaves develop.

Identification

a. stigma b. style c. ovary d. ovule e. sepal f. anther g. filament h. petal i. receptacle j. pedicel

ANSWERS TO THE SUMMARY OF MODULE #16

1. Reptiles have the following six characteristics in common: <u>covered with tough dry scales</u>, <u>ectothermic</u>, <u>breathe with lungs throughout their lives</u>, <u>three-chambered heart with a ventricle that is partially divided</u>, <u>produce amniotic eggs covered with a leathery shell</u>, <u>most oviparous but some ovoviviparous</u>.

2. Scales prevent <u>water loss</u> and <u>insulate</u>. Since scales are not made of living tissue, reptiles must <u>molt</u> in order to grow.

3. An amniotic egg is covered in a protective <u>shell</u>. The <u>amnion</u> grows around the embryo, forming a fluid-filled sac in which the embryo floats. The <u>yolk sac</u> contains the <u>yolk</u>, which feeds the embryo. The <u>allantois</u> is a sac of blood vessels that allows for the respiration and excretion of the embryo. The <u>chorion</u> is a membrane that envelopes these four structures. The <u>albumen</u> protects the embryo from infection, is a storehouse for water, provides mechanical support for the chorion, and stores proteins, which can be broken down into <u>amino acids</u> that the embryo can use in <u>biosynthesis</u>.

4. The <u>tuatara</u> is in order Rhynchocephalia, which also contains many extinct reptiles.

5. Lizards and snakes are in order <u>Squamata</u>. Lizards have two pairs of limbs, while snakes have <u>none</u>. Lizards have ears and can hear, while snakes are <u>deaf</u>. Lizards have the same type of scales all over their bodies, while snakes have specialized scales on their <u>bellies</u> for locomotion. Most lizards have eyelids and can therefore close their eyes, but snakes <u>cannot</u>.

6. A snake has two nostrils for smelling, but it augments this sense with sensory pits called <u>Jacobson's organs</u>. When a snake sticks out its tongue, it collects <u>chemicals suspended in the air</u>. It then pulls its tongue back into its mouth, transferring them to the <u>Jacobson's organs</u>, which send signals to the brain. Poisonous snakes generally produce <u>neurotoxins</u> (which are fast-acting) or <u>hemotoxins</u> (which take longer to do damage but are more deadly). <u>Pit vipers</u> are poisonous snakes that have two heat-sensing pits which allow them to sense <u>warm-blooded creatures</u>, even if there is no light.

7. Turtles and tortoises are found in order <u>Testudines</u>. If it lives in the water, it is a <u>turtle</u>, but if it lives on land, it is a <u>tortoise</u>.

8. Alligators and crocodiles make up order <u>Crocodilia</u>. When an <u>alligator's</u> mouth is closed, you cannot see its teeth. <u>Crocodile</u> teeth can be seen even when the animal's mouth is closed.

9. Although a few reptiles that are classified as dinosaurs (like the tuatara) are still living, most dinosaurs <u>are extinct</u>. Although we know little about these creatures, their fossils do tell us some things. Apatosaurus was one of the <u>sauropods</u>, which were probably the largest dinosaurs. Stegosaurus was one of the <u>thyreophorans</u>, which were probably herbivores. Triceratops was one of the <u>marginocephalia</u>, which means "fringe heads." Tyrannosaurus was one of the <u>theropods</u>, which stood on hind legs and ranged from very small to very large. Plesiosaurus was one of the <u>sauropterygia</u>, which were technically not dinosaurs, but were marine reptiles with flippers. Pteranodon was also not technically a dinosaur. Instead, it was a large <u>flying reptile</u>.

10. Birds (members of class <u>Aves</u>) have the following six characteristics in common: <u>endothermic</u>, <u>heart with four chambers</u>, <u>toothless bill</u>, <u>oviparous laying an amniotic egg that is covered in a lime-containing shell</u>, <u>covered with feathers</u>, and <u>skeleton composed of porous, lightweight bones</u>.

11. The three most important features that allow birds to fly are <u>feathers</u>, <u>wings</u>, and <u>skeletal structure</u>. Feathers are composed of a <u>shaft</u> from which <u>barbs</u> extend. Contour feathers (used for flight) have interlocking <u>smooth barbules</u> and <u>hooked barbules</u>, while down feathers (used for insulation) have only <u>smooth barbules</u>. Because feathers are not made of living tissue, they must be <u>molted</u>, but the process occurs very precisely so that the bird is never out of <u>balance</u>. The wings are equipped with turbulence-dampening devices called <u>alulas</u> that were copied so that airplane wings would not fall off. Birds that fly have <u>porous</u> bones that are lightweight but strong, because they are reinforced with <u>struts</u> that were also copied in aircraft design.

12. Mammals have the following five characteristics in common: <u>hair covering the skin</u>, <u>internal fertilization and usually viviparous</u>, <u>nourish their young with milk</u>, <u>four-chambered heart</u>, <u>endothermic</u>.

13. Mammal hair comes in two forms: <u>guard hair</u> on top of <u>underhair</u>. The latter serves as insulation.

14. Since most mammals are viviparous, the embryo is attached to the mother by a <u>placenta</u>. The - <u>nonplacental</u> mammals, however, are either <u>oviparous</u>, or their embryos develop in a <u>pouch</u>. The length of the <u>gestation period</u> determines how well developed the offspring is when it is born.

15. Female mammals secrete milk to nourish their young from special glands called <u>mammary glands</u>.

Identification

a. <u>amniotic fluid</u> b. <u>embryo</u> c. <u>amnion</u> d. <u>allantois</u> e. <u>chorion</u> f. <u>yolk sac</u> g. <u>yolk</u> h. <u>albumen</u> i. <u>shell</u>

TEST FOR MODULE #1

1. Give definitions for the following terms:

   a. Heterotrophs
   b. Mutation
   c. Theory
   d. Photosynthesis
   e. Prokaryotic cell
   f. Species

2. An organism is classified as a decomposer. Is it an autotroph or a heterotroph?

3. A tiger develops lockjaw and can no longer open its mouth. Which of the four life functions will it not be able to perform?

4. In laboratory studies, organisms from group A produce offspring with subtle differences as compared to the parents. Organisms from group B produce offspring that are identical to the parents, and organisms from group C produce offspring with marked differences as compared to the parents. Which group reproduces asexually? Which reproduces sexually? Which group is experiencing mutations?

5. A biologist studies two organisms from the same family. She then studies two organisms from the same genus in that family. In which case do you expect the most similarity between organisms?

6. A biologist wishes to study how two remarkably different plants react to the same changes in their surroundings. Should he choose organisms from different phyla or different orders within kingdom Plantae?

7. Why did the law of spontaneous generation survive for so many years?

8. Classify the following organisms using the biological key in the appendix:

a.

This organism is green.

b.

Photos from www.clipart.com

9. An organism is a multicellular decomposer with eukaryotic cells. To what kingdom does it belong?

10. An organism is a multicellular autotroph with eukaryotic cells. To what kingdom does it belong?

11. An organism has a single, prokaryotic cell. To what kingdom does it belong?

12. If we were using the three-domain system of classification, in which domain would an organism made of a single, eukaryotic cell be classified?

TEST FOR MODULE #2

1. Define the following terms:

    a. Logistic growth
    b. Plasmid
    c. Endospore

2-5. Identify the numbered structures in the diagram below.

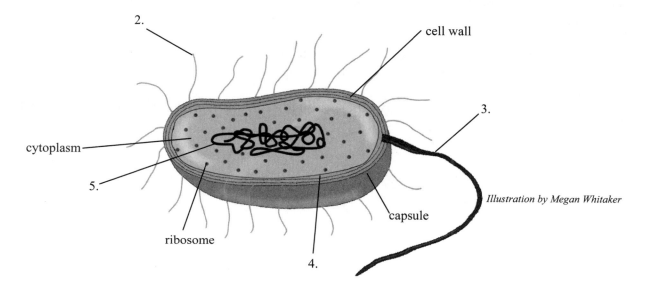

Illustration by Megan Whitaker

6. A bacterium suddenly cannot manufacture proteins. What component(s) of the cell is (are) not working?

7. If a bacterium is anaerobic, would you expect to find it floating at the top of a lake or deep in the muck at the bottom of the lake?

8. A bacterium is photosynthetic. Is it a decomposer?

9. A bacterium receives a new trait that it did not previously have. However, it did not participate in conjugation. How is this possible?

10. What bacterial process does the following schematic represent?

11. If a sample of food is dehydrated, what condition for bacterial growth are you removing from the food?

12. If a population of bacteria is in steady state, does that mean no bacteria are dying?

13. What shape is a bacterium from the genus *Streptobacillus*.

14. A population is composed of bacteria that are very sensitive to light. They live in a lake that is in a cave, so they flourish. Slowly, however, a hole begins to erode in the cave's roof. As the days pass, light begins to filter in and the cave starts to be dimly lit. The bacteria begin to die. While the erosion is taking place, however, two individual bacteria in the soil above ground fall into the lake. These bacteria cannot survive in the dimness of the poorly-lit cave, so they die immediately. Days later, the hole has opened up so that light floods the cavern. Nevertheless, the bacteria that were once dying are flourishing, with a population larger than they had before the erosion. Explain how this could have happened.

15. A bacterium is Gram negative and needs light to survive. It lives in a habitat that is completely oxygen-free. To what phylum and class does it belong?

16. A Staphylococcus bacterium is Gram positive and lives in an oxygen-free habitat. To what phylum and class does it belong?

TEST FOR MODULE #3

The classification groups within kingdom Protista

Subkingdoms	Phyla (in no particular order)
Algae	Mastigophora
Protozoa	Sarcodina
	Chlorophyta
	Chrysophyta
	Rhodophyta
	Sporozoa
	Pyrrophyta
	Ciliophora
	Phaeophyta

1. Define the following terms:

   a. Thallus
   b. Symbiosis
   c. Vacuole

2. If an organism from kingdom Protista is heterotrophic, what subkingdom is it most likely in?

3. What phylum produces organisms whose remains are used as an abrasive in toothpaste? What is the generic name given to these organisms?

4. Some forms of algae have a chemical that can be used as a thickening agent in many consumer products. What is this chemical and in what phylum do these algae belong?

5. What phylum contains organisms that must have two nuclei? What are these nuclei called, and what is the function of each?

6. Name one organism in kingdom Protista that is pathogenic. Name the malady that this organism causes.

7. If all diatoms were to suddenly go extinct, what would happen to the earth's atmosphere?

8. An organism forms a hard shell around itself in response to life-threatening conditions. If those life-threatening conditions had not occurred, it never would have behaved in such a way. Is this organism from phylum Sporozoa? Why or why not?

9. Two samples of cytoplasm from an amoeba are studied. The first is thin and watery, while the second is thick. Which sample was taken near the plasma membrane, and which was taken from the center of the amoeba?

10. An organism from phylum Mastigophora cannot move. What organelle is not functioning?

11. An organism from phylum Ciliophora has no place to store its food once it intakes the food. What organelle is it missing?

12. If an organism in subkingdom Algae has chlorophyll, what organelle must it also have?

13. Even though kingdom Protista is mostly known for its single-celled creatures, there are two phyla that contain multicellular organisms. What phyla are they?

14. Classify each organism below into its proper subkingdom and phylum. Remember that there is a list of the subkingdoms and phyla on the previous page.

Photos by Kathleen J. Wile

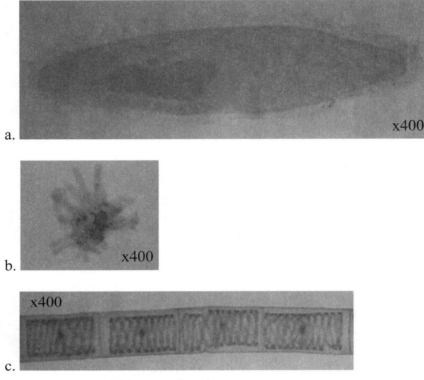

a.

b.

c.

This organism is green.

TEST FOR MODULE #4

1. Define the following terms:

    a. Extracellular digestion
    b. Rhizoid hypha
    c. Stolon
    d. Fermentation
    e. Hypha

2. If a fungus forms haustoria, is it saprophytic or parasitic?

3. What kind of hypha exists in all multicellular fungi?

4. What does chitin provide for a fungus?

5. Name two specialized aerial hyphae.

6. Given the phyla of kingdom Fungi: Basidiomycota, Ascomycota, Zygomycota, Chytridiomycota, Deuteromycota, and Myxomycota, classify fungi with these characteristics:

    a. form spores on clublike basidia
    b. have no known sexual mode of reproduction
    c. form sexual spores where hyphae fuse
    d. resemble both protozoa and fungi

7. What part of the mushroom (the stipe, cap, or gill) holds the basidia?

8. Give the means of sexual reproduction and one means of asexual reproduction employed by bread molds.

9. Name two pathogenic fungi and the maladies that they cause.

10. A farmer notices that a certain crop grows much better in the presence of a certain fungi. What is the most likely explanation?

11. What useful medicine is produced by fungi in genus *Penicillium*? There is a general name for such medicines. What is that general name?

12. A biologist looks through a microscope at a single-celled life form. The microscope is not good enough to discern whether the cell is eukaryotic or prokaryotic. However, the biologist does see that the cell reproduces by budding. What is the most likely kingdom in which this organism belongs?

13. Why do slime molds appear in kingdom Protista in some biology books?

TEST FOR MODULE #5

1. Define the following terms:

    a. Saturated fat
    b. Physical change
    c. Model
    d. Isomers

2. A student reports on an atom that has 11 protons, 10 neutrons, and 10 electrons. What is wrong with the student's report?

3. Two atoms have the same number of neutrons but different numbers of protons. Do they belong to the same element?

4. The element sulfur is comprised of all atoms that have 16 protons. How many neutrons are in sulfur-34?

5. How many total atoms are in one molecule of $C_{22}H_{44}O$?

6. Identify the following as an atom, element, or molecule:

    a. P   b. N_2 c. oxygen-16   d. PH_3

7. When you boil water, are you causing a chemical change or a physical change?

8. Two solutions of different solute concentration are separated by a membrane. After 1 hour, the solutions are of equal concentrations and the water levels are the same as when the experiment started. Is this an example of diffusion or osmosis?

9. In the following equation:

$$C_3H_8 + 5O_2 \rightarrow 3CO_2 + 4H_2O$$

Is CO_2 a product or a reactant? How many molecules of it are involved in the reaction?

10. What four things are necessary for photosynthesis?

11. If the following molecule is a carbohydrate, what is the value of x?

$$C_7H_xO_2$$

12. If you perform hydrolysis on a disaccharide, what kind of molecules will you get?

13. The pH of three solutions is measured. Solution A has a pH of 1.5; the pH of solution B is 7.1; and solution C has a pH of 13.2. Which solution or solutions is or are acidic?

14. Which of the following is an unsaturated fatty acid molecule?

a.   b.   c.

15. What determines the properties of a protein?

16. What part of the nucleotide is responsible for the way DNA stores its information?

17. What holds the two helixes in DNA together?

TEST FOR MODULE #6

1. Define the following terms:

    a. Cytoplasmic streaming
    b. Isotonic solution
    c. Cytolysis
    d. Phagocytosis
    e. Activation energy
    f. Plasmolysis

2. A cell produces a protein that will be used by other cells. When it ejects the protein, has it performed egestion, secretion, or excretion?

3. What lies between the cell walls of a plant's cells?

4. What is the difference between digestion and respiration?

5. Name at least two organelles that are involved in biosynthesis.

6. What organelle does rough ER have which smooth ER does not have?

7. Is a chloroplast an example of a leucoplast or a chromoplast?

8. Before a polysaccharide can be used in cellular respiration in an animal cell, to what organelle must it be sent?

9. If a cell's mitochondria stop working, can it perform any cellular respiration?

10. Which provides more energy per molecule of glucose: respiration in aerobic conditions or respiration in anaerobic conditions?

11. What stage in cellular respiration produces the most energy?

12. What gives the plasma membrane the ability to self-reassemble?

13. Which two stages in aerobic cellular respiration produce equal amounts of ATP?

The test continues on the next page.

14. Name the structures identified by letters in the figure below:

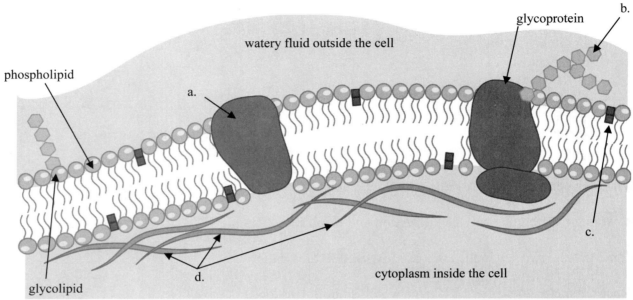

Illustration by Speartoons

TEST FOR MODULE #7

1. Define the following terms:

    a. Interphase
    b. Karyotype
    c. Diploid cell
    d. Gametes
    e. Virus

2. What factors besides genetics play a role in determining the characteristics of a person?

3. Identify the stage of mitosis represented by each picture, and then list the stages in the proper order.

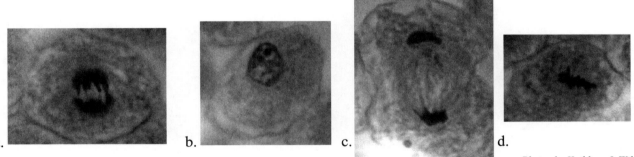

a.   b.   c.   d.

Photos by Kathleen J. Wile

4. A cell's DNA consists of 12 pairs of homologous chromosomes. What is its diploid number? What is its haploid number?

5. Three cells that each has a diploid number of 32 go through mitosis. How many cells result and how many total chromosomes are in each cell?

6. Three cells that each has a diploid number of 32 go through meiosis. How many cells result and how many total chromosomes are in each cell?

7. Which resembles mitosis most: meiosis I or meiosis II?

8. A haploid cell with duplicated chromosomes turns into two haploid cells with no duplicated chromosomes. Did the cell go through mitosis, meiosis I, or meiosis II?

9. A diploid cell with duplicated chromosomes turns into two diploid cells with no duplicated chromosomes. Did the cell go through mitosis, meiosis I, or meiosis II?

10. A single diploid cell goes through meiosis. Only one useful gamete is produced. Did this meiosis take place in a male or female?

11. What does a vaccine do to make a person immune to a virus?

12. What two different kinds of genetic material are found in viruses?

13. A strand of tRNA has the following nucleotide sequence:

adenine, uracil, guanine

What codon in mRNA attracts this strand of tRNA?

14. Part of an mRNA strand has the following nucleotide sequence:

uracil, guanine, cytosine, cytosine, guanine, adenine, uracil, adenine, adenine

For how many amino acids does this part of the strand code?

15. What DNA sequence produced the mRNA sequence given in problem #14?

TEST FOR MODULE #8

1. Define the following terms:

    a. True breeding
    b. Allele
    c. Recessive allele
    d. Monohybrid cross
    e. Dihybrid cross
    f. Autosomal inheritance

2. State the principles of Mendelian genetics using the updated terminology.

3. In humans, the ability to roll one's tongue is a dominant genetic trait. If "R" represents this allele and "r" represents the recessive allele, what are the possible genotypes for a person who can roll his tongue?

4. For a given trait, how many alleles does a normal gamete have?

5. For a given trait, how many alleles does a non-gamete cell have?

6. a. The ability for a person to taste PTC is a dominant genetic trait ("T"), while the inability to taste PTC is recessive ("t"). If a man is heterozygous in that trait, what is his genotype?

    b. If a woman cannot taste PTC, what is her genotype?

    c. Determine the percentage chance of each genotype for the children of the man in (a) and the woman in (b).

7. Hemophilia is a sex-linked, recessive trait.

    a. Write the Punnett square for a non-hemophilic man having children with a woman who carries but does not have the disease.

    b. What percentage of the girls will have the disease?

    c. What percentage of the boys will have the disease?

8. Why do recessive phenotypes in sex-linked traits show up in males significantly more often than in females?

9. Why would you not expect twins (who have identical DNA) to be identical in every way?

10. Name four means by which genetic disorders arise.

Test continues on the next page.

11. A pea plant is heterozygous in both the color of the pea produced ("Y" for yellow and "y" for green) and the height of the plant ("T" for tall and "t" for short).

a. What possible combinations of alleles exist in its gametes?

b. Draw the resulting Punnett square for when this plant is self-pollinated.

c. What are the possible *phenotypes* and their percent chances?

12. The following pedigree is for humans, concentrating on the ability to roll their tongues. Since you already know that the ability to roll your tongue is dominant and the inability to do so is recessive, determine whether the filled circles and squares represent those who can or those who cannot roll their tongues.

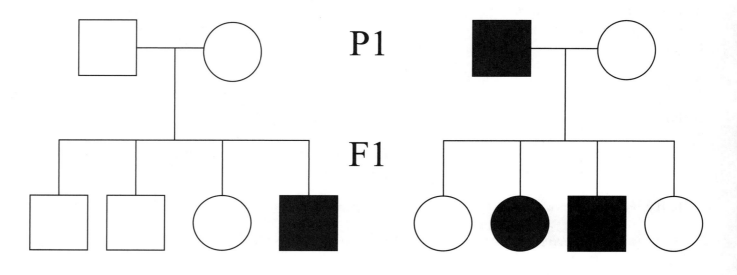

TEST FOR MODULE #9

1. Define the following terms:

    a. The immutability of species
    b. Microevolution
    c. Macroevolution
    d. Strata
    e. Fossils
    f. Structural Homology

2. What ship did Darwin travel on while working on his research?

3. Name two ideas from other scientists that influenced Darwin in his work.

4. A horse breeder takes a horse that has won several races and breeds it with another horse that has won several races. These two horses produce 3 colts over their lifetime, and the breeder takes the fastest of these horses and breeds them again. After several generations of such a process, the breeder has produced a horse faster than any other in the world. Is this an example of microevolution or macroevolution?

5. If a colony of bacteria were to give rise to an amoeba, would this be an example of microevolution or macroevolution?

6. Of the four basic data sets discussed in this module, which ones provide conclusive evidence for macroevolution?

7. Of the four basic data sets discussed in this module, which ones provide conclusive evidence against macroevolution?

8. Some creationists say that all of the ideas set forth in Darwin's book, *The Origin of Species*, are wrong. Why is this not true?

9. The sequence of amino acids in the hemoglobin of a human is determined and compared to the sequence of amino acids in the hemoglobin of a rat and an ape. According to macroevolutionists, which of the two amino acids sequences should be closer to that of the human's?

10. How does the neo-Darwinist hypothesis differ from the hypothesis of Darwin?

11. What problem with both Darwin's original hypothesis and neo-Darwinism does punctuated equilibrium attempt to solve?

12. Bacteria can sometimes become immune to the effects of an antibiotic. Is mutation the only way this can happen?

13. Fill in the blanks: The fact that representatives of all major animal phyla can be found in some of the lowest sedimentary rock in the geological column is often referred to as the _____ _____.

14. Consider the following amino acid sequences that make up a small portion of a protein:

a. Gly-Ile-Gly-Gly-Arg-His-Gly-Gly-Glu(NH$_2$)-Glu-Glu(NH$_2$)-Lys-Lys-Lys

b. Gly-Leu-Phe-Gly-Arg-Lys-Ser-Gly-Glu(NH$_2$)-Gly-Glu(NH$_2$)-Ala-Arg-Lys

c. Leu-Ile-Gly-Gly-Arg-His-Ser-Gly-Glu(NH$_2$)-Ala-Glu(NH$_2$)-Arg-Arg-Arg

Which protein would you expect to be the most similar to a protein with the following subset of amino acids?

Phe-Ile-Gly-Gly-Arg-His-Gly-Gly-Glu(NH$_2$)-Glu-Glu(NH$_2$)-Lys-Lys-Lys

TEST FOR MODULE #10

1. Define the following terms:

    a. Ecosystem
    b. Biomass
    c. Watershed
    d. Transpiration
    e. Greenhouse effect

<u>Questions 2-4 refer to the following food web diagram</u>:

Illustrations from www.clipart.com

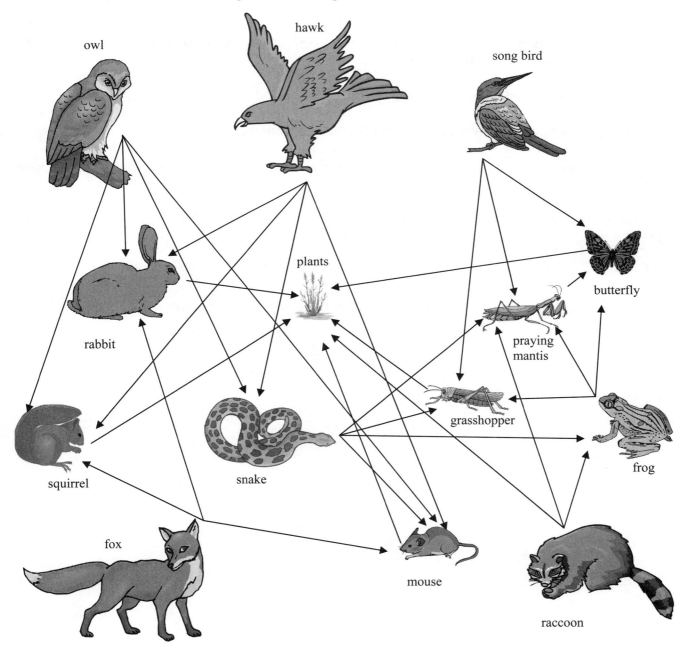

owl

hawk

song bird

plants

butterfly

rabbit

praying
mantis

squirrel

snake

grasshopper

frog

fox

mouse

raccoon

2. What is the most likely consequence of removing the snake and raccoon from this ecosystem?

3. What are the possible trophic levels of the owl?

4. What are the possible trophic levels of the raccoon?

5. Draw an ecological pyramid of an ecosystem that has twice as much biomass in producers as there is in primary consumers, twice as much biomass in primary consumers as there is in secondary consumers, and twice as much biomass in secondary consumers as there is in tertiary consumers.

6. Name two of the mutualistic relationships that we learned about in this module and briefly discuss the role of each participant.

7. The stream flowing from a watershed has an algal bloom caused by too many nutrients in the water. What is the most likely explanation for this?

8. Does photosynthesis add oxygen to the air or remove it?

9. Besides photosynthesis, how is carbon dioxide removed from the air?

10. If surface runoff did not occur in the water cycle of an ocean shore ecosystem, what would happen to the ocean?

11. Explain this statement: "The greenhouse effect is a good thing, but global warming is too much of a good thing."

12. Is global warming occurring today?

13. What do nitrogen-fixing bacteria provide to other organisms? Do they provide it directly to producers or consumers?

TEST FOR MODULE #11

1. Define the following terms:

   a. Invertebrates
   b. Vertebrates
   c. Nematocysts
   d. Posterior end
   e. Hermaphroditic
   f. Mantle

2. Place each of the following organisms into its proper phylum (Porifera, Cnidaria, Annelida, Mollusca, Platyhelminthes):

    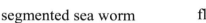

a.   b.   c.   d.   e.

sponge   snail   jellyfish   segmented sea worm   flatworm

a. Illustration from www.clipart.com
b-d. Illustrations from the MasterClips collection
e. Illustration by Kathleen J. Wile

3. Are the following organisms univalve or bivalve mollusks?

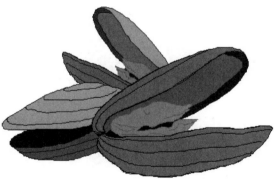

Illustration from the MasterClips collection

4. What do sponges eat and how do they get their food?

5. What is the difference between a hard, prickly sponge and a soft sponge?

6. Which phylum of animals discussed in this module has nematocysts?

7. An earthworm's seminal vesicles are full. Has it mated yet?

8. What function does the cuticle perform in an earthworm?

9. Two members of phylum Platyhelminthes are studied. The first has a complex nervous system and the second does not. Which one is most likely the parasitic planarian?

10. Identify structures a-g in the following figure:

Illustration by Megan Whitaker

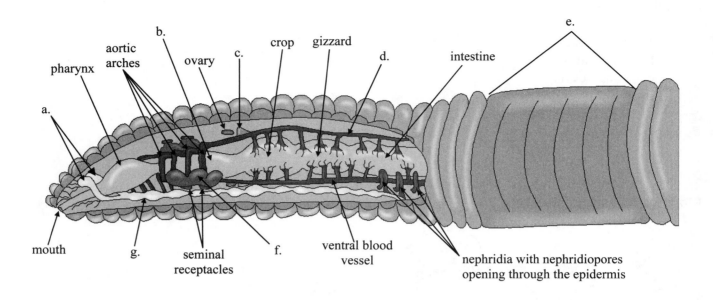

11. What is the common mode of asexual reproduction among the organisms of phylum Cnidaria?

12. What is the common mode of asexual reproduction among the organisms of phylum Platyhelminthes?

TEST FOR MODULE #12

1. Define the following terms:

    a. Exoskeleton
    b. Simple eye
    c. Open circulatory system
    d. Statocyst
    e. Gonad

2. Identify structures a-c in the following photograph:

Photo by Kathleen J. Wile

3. Identify structures a-g in the following diagram:

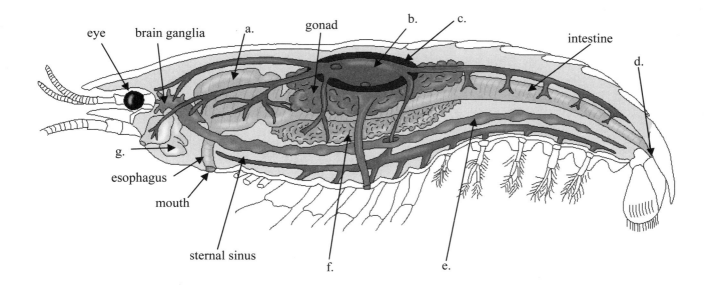

4. Why do arthropods molt?

5. What five characteristics set arachnids apart from the other arthropods?

6. What four characteristics set insects apart from the other arthropods?

7. What are the stages of incomplete metamorphosis?

8. What takes the place of a respiratory system in insects?

9. If an insect has horny wings that protect membranous wings, to what order does it belong?

10. If an insect has leather-like wings that protect membranous wings, to what order does it belong?

TEST FOR MODULE #13

1. Define the following terms:

    a. Bone marrow
    b. Cerebrum
    c. Ectothermic

2. Name at least two chordates that have a larval stage as a part of their life cycles.

3. Assign the following chordates to one of these classifications: subphylum Urochordata, subphylum Cephalochordata, class Agnatha, class Chondrichthyes, class Osteichthyes, or class Amphibia.

a. Ray   b. Salmon   c. Sea squirt   d. Toad   e. Salamander   f. Lancelet   g. Lamprey eel

4. If an animal's optic lobes are very small, what, most likely, is its weakest sense?

5. What protein allows red blood cells to perform their function?

6. An animal reproduces when the female lays eggs and the male then fertilizes them. The eggs are then left to develop and hatch. Is fertilization external or internal? What kind of development is this?

7. Which has the most flexible skeleton, a shark, a frog, or a lionfish?

8. Identify structures a-d in the following diagram:

Illustration by Megan Whitaker

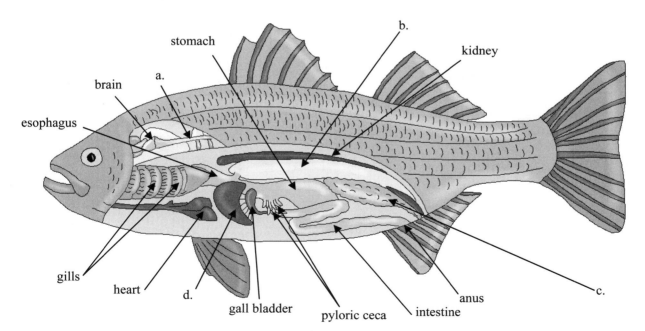

9. Identify the functions of structures a-d in problem #8

10. For each arrow in the figure below, indicate whether it is pointing to a vein or an artery.

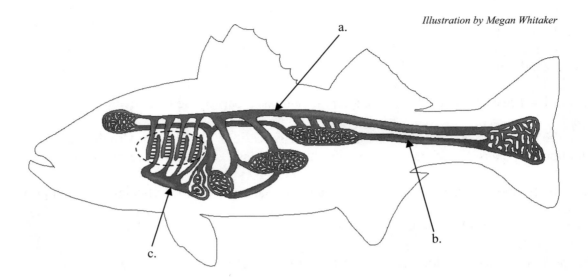

Illustration by Megan Whitaker

11. For each arrow in problem #10, indicate whether the blood is oxygen poor or oxygen rich.

12. What is the major respiratory organ for most amphibians?

TEST FOR MODULE #14

1. Define the following terms:

    a. Vegetative organs
    b. Reproductive plant organs
    c. Undifferentiated cells
    d. Phloem
    e. Deciduous plant

2. Determine the shape, margin, and venation of the following leaves:

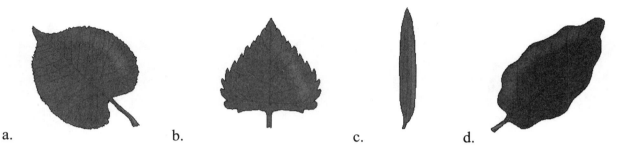

a.   b.   c.   d.

Illustrations by Megan Whitaker

3. What function do the guard cells perform in a leaf?

4. A leaf has the spongy mesophyll on top. Which side of the leaf (top or bottom) will be the lighter shade of green?

5. What structure in a deciduous tree causes the leaves to die and fall off in the autumn?

6. What are anthocyanins, and what do they do to a leaf's color?

7. In which region of a root do you find undifferentiated cells?

8. Is this fibrovascular bundle from a monocot or a dicot?

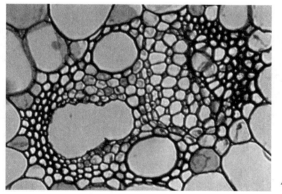

Photo by Kathleen J. Wile

9. Why is the bark of a tree often cracked?

10. A tree has seed cones and pollen cones. To which phylum does it belong?

11. A plant has a primary root that grows and grows without branching. What kind of root system is this?

12. Name two differences between monocots and dicots.

TEST FOR MODULE #15

1. Define the following terms:

    a. Physiology
    b. Nastic movement
    c. Pore spaces
    d. Translocation
    e. Gravitropism
    f. Imperfect flowers

2. Name the four processes for which plants require water.

3. If a plant loses control of its stomata and they remain closed, will substances still flow through its xylem?

4. If a plant loses control of its stomata and they remain closed, will substances still flow through its phloem?

5. A biologist has a sample of fluid that came from a plant. If it is composed mostly of organic materials, did it come from the plant's xylem or phloem?

6. If an insectivorous plant cannot catch any insects, will it starve to death?

7. A gardener has two genetically identical plants. If one is the offspring of the other, did the parent propagate sexually or vegetatively?

8. Identify structures a-d in the figure below:

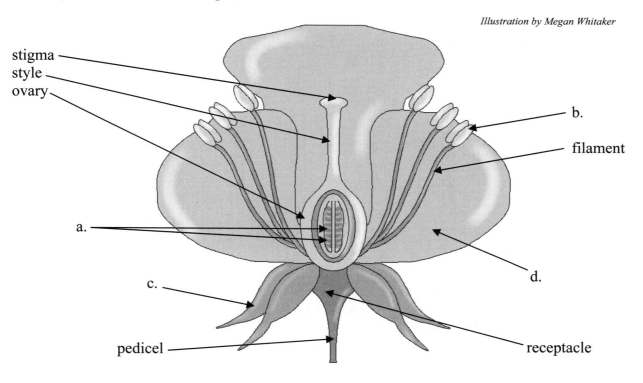

Illustration by Megan Whitaker

stigma
style
ovary

b.

filament

a.

d.

c.

pedicel

receptacle

Match the following structures with the correct function.

9. Anther
10. Ovule
11. Stigma
12. Sepal

a. Protects the sexual organs as they form
b. Forms and releases pollen grains
c. Holds the embryo sac
d. Catches pollen grains

13. What function do cotyledons perform before germination?

14. What is the purpose of a fruit?

15. Name at least three ways in which pollen is transferred from the stamens of one flower to the carpels of another.

16. Which has more cells prior to fertilization, a pollen grain or an embryo sac?

17. When an embryo sac is fertilized, which cell is not diploid: the zygote or the endosperm?

TEST FOR MODULE #16

1. Define the following terms:

    a. Amniotic egg
    b. Hemotoxin
    c. Endotherm
    d. Placenta
    e. Gestation

2. A vertebrate has a four-chambered heart, lays amniotic eggs with a lime-containing shell, and it has a light, porous skeleton. Is it a reptile, bird, or mammal?

3. A vertebrate has dry, tough scales, a three-chambered heart with a partial division in the ventricle, and it breathes with lungs. Is it a reptile, bird, or mammal?

4. A vertebrate has a four-chambered heart, hair, and nourishes its young with its own milk. Is it a reptile, bird, or mammal?

5. These are the reptile orders that contain currently living reptiles:

Crocodilia, Testudines, Rhynchocephalia, Squamata

Place the following types of reptiles into their appropriate order:

a. tuataras   b. lizards   c. crocodiles   d. turtles

6. A living vertebrate is endothermic. Is it possible that the vertebrate is a reptile?

7. Which kind of feather has hooked barbules: contour feathers or down feathers?

8. You find an egg in your back yard. It is covered in a soft, leathery shell. Is it a reptile egg or a bird egg?

9. A bird's feathers become inflexible because the hooked barbules do not slide easily on the smooth barbules. What should the bird do to fix this problem?

10. A mammal lives in a cold climate and needs a lot of insulation. Will its guard hair or underhair be significantly thicker than the average mammal's?

11. Two species of animal are very similar in appearance.   The first species gives birth to offspring that have no hair. Their eyes are also closed. The second species gives birth to offspring that have a full coat of hair. Their eyes are also open. Which species has the shorter gestation period?

The test continues on the next page.

12. Identify the structures in the amniotic egg pictured below:

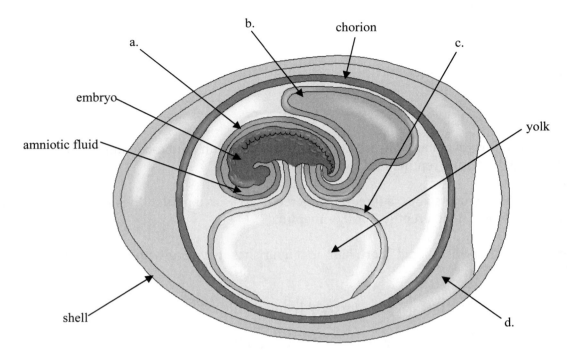

Illustration by Megan Whitaker

SOLUTIONS TO THE TEST FOR MODULE #1

1. (6 pts – one for each definition)

a. <u>Heterotrophs</u> – Organisms that depend on other organisms for their food

b. <u>Mutation</u> – An abrupt and marked change in the DNA of an organism compared to that of its parents

c. <u>Theory</u> – A hypothesis that has been tested with a significant amount of data

d. <u>Photosynthesis</u> – The process by which green plants and some other organisms use the energy of sunlight and simple chemicals to produce their own food

e. <u>Prokaryotic cell</u> – A cell that has no distinct, membrane-bounded organelles

f. <u>Species</u> – A unit of one or more populations of individuals that can reproduce under normal conditions, produce fertile offspring, and are reproductively isolated from other such units

2. (1 pt) It is a <u>heterotroph</u>, because autotrophs make their own food.

3. (1 pt) <u>It will not be able to extract energy from the surroundings and convert it into energy that sustains life</u>, because it will be unable to eat.

4. (2 pts – 1 pt for one correct answer, ½ pt for each correct answer after that.) <u>Group A reproduce sexually</u> because in sexual reproduction, the offspring have slight differences from the parents. <u>Group B reproduces asexually</u> because asexual reproduction leads to offspring genetically identical to the parent. <u>Group C is experiencing mutations</u>, because mutations lead to abrupt and marked differences between offspring and parent or parents.

5. (1 pt) <u>The two organisms from the same genus will probably have the most similarity</u>, as a genus contains fewer species than a family.

6. (1 pt) He should choose organisms from <u>different phyla</u>, as the differences between organisms within a group are more pronounced the higher the group is in the classification scheme.

7. (1 pt) <u>It survived because flawed experiments seemed to confirmed it</u>.

8. (3 pts – ½ point for each proper classification)
a. Kingdom: <u>Plantae</u>   Phylum: <u>Anthophyta</u>   Class: <u>Dicotyledoneae</u>
b. Kingdom: <u>Animalia</u> Phylum: <u>Chordata</u>   Class: <u>Aves</u>

9. (1 pt) Multicellular decomposers are mostly in kingdom <u>Fungi</u>.

10. (1 pt) Multicellular, eukaryotic organisms that make their own food are in kingdom <u>Plantae</u>.

11. (1 pt) Organisms with prokaryotic cells are in kingdom <u>Monera</u>.

12. (1 pt) In the three-domain system, all eukaryotic organisms belong in domain <u>Eukarya</u>.

Total possible points: 20

SOLUTIONS TO THE TEST FOR MODULE #2

1. (3 pts – one for each definition)

a. <u>Logistic growth</u> – Population growth that is controlled by limited resources

b. <u>Plasmid</u> – A small, circular section of extra DNA that confers one or more traits to a bacterium and can be reproduced separately from the main bacterial genetic code

c. <u>Endospore</u> – The DNA and other essential parts of a bacterium coated with several hard layers

2. (1 pt) <u>Fimbria</u>

3. (1 pt) <u>Flagellum</u>

4. (1 pt) <u>Plasma membrane</u>

5. (1 pt) <u>DNA</u>

6. (1 pt) Since <u>the ribosomes</u> are where proteins are manufactured in the cell, they must not be working.

7. (1 pt) You would expect to find it <u>deep in the muck</u>, because there is little or no oxygen down there.

8. (1 pt) <u>It is not a decomposer</u>, because photosynthetic bacteria are autotrophs.

9. (1 pt) <u>It participated in transformation</u>, one of the other means by which bacteria get new traits. The student could also say "transduction" here. That would also receive full credit.

10. (1 pt) It represents <u>asexual reproduction</u>.

11. (1 pt) <u>Moisture</u> is being removed.

12. (1 pt) <u>A steady state does not mean that no bacteria are dying</u>. In a steady state, the population does not change because bacteria are reproducing as quickly as others are dying.

13. (1 pt) Since the name ends in bacillus, it is <u>rod-shaped</u>. You could also say bacillus.

14. (2 pts – one point for mentioning transformation, and one for mentioning either conjugation or asexual reproduction) <u>A bacterium must have participated in transformation</u> so that the ability to live in light was absorbed from one or both of the dead bacteria that fell into the lake. Then, because there are so many bacteria present, <u>that bacterium must have participated in conjugation</u> with several others. NOTE: After mentioning transformation, the student could say that the bacterium then started asexually reproducing to rebuild the population. That is acceptable in place of conjugation.

15. (2 pts – one for the phylum and one for the class) Gram negative bacteria are in <u>phylum Gracilicutes</u>. Anaerobic bacteria in this phylum are in <u>class Anoxyphotobacteria</u>.

16. (2 pts – one for the phylum and one for the class) Gram positive bacteria are in <u>phylum Firmicutes</u>. <u>Class Firmibacteria</u> contains those Gram positive bacteria that are coccus-shaped. The oxygen-free information is irrelevant, because respiration only matters in classifying Gram-negative bacteria.

Total possible points: 21

SOLUTIONS TO THE TEST FOR MODULE #3

1. (3 pts – one for each definition)

a. <u>Thallus</u> – The body of a plant-like organism that is not divided into leaves, roots, or stems

b. <u>Symbiosis</u> – A close relationship between two or more species where at least one benefits

c. <u>Vacuole</u> – A membrane-bounded "sac" within a cell

2. (1 pt) It is most likely in subkingdom <u>Protozoa</u>, since subkingdom Algae holds autotrophic organisms.

3. (2 pts – one for the phylum and one for the generic name) These organisms come from <u>phylum Chrysophyta</u> and are commonly called <u>diatoms</u>.

4. (2 pts – one for the chemical and one for the phylum) The chemical is <u>alginic acid</u> (the student could say algin), and it comes from members of phylum <u>Phaeophyta</u>.

5. (2 pts – one for the phylum, ¼ for each name, ¼ for each function) Members of phylum <u>Ciliophora</u> have two nuclei, the <u>macronucleus</u> and the <u>micronucleus</u>. <u>The macronucleus controls the organism's metabolism</u>, while <u>the micronucleus controls reproduction</u>.

6. (1 pt – ½ for the organism, ½ for the malady) There are several. Any of the following are correct answers:

<u>Entamoeba histolytica causes dysentery</u>
<u>Trypanosoma causes African sleeping sickness</u>
<u>Balantidium coli causes dysentery</u>
<u>Plasmodium causes malaria</u>
<u>Toxoplasma causes toxoplasmosis</u>

7. (1 pt) <u>The oxygen supply in the atmosphere would dwindle away</u>, because diatoms perform a large amount of the photosynthesis which replaces the oxygen that most organisms use.

8. (2 pts – one for saying it is not in the phylum, one for why.) <u>It is not from phylum Sporozoa</u> because <u>members of this phylum form spores as a natural part of their life cycle</u>. Whether or not conditions are threatening does not influence the production of spores in this phylum.

9. (1 pt) <u>The first was taken near the plasma membrane</u> because ectoplasm is thin and watery and near the plasma membrane. <u>The second was taken near the center of the amoeba</u> because endoplasm is thicker and nearer to the center of the amoeba.

10. (1 pt) Since members of phylum Mastigophora use flagella to move, <u>the flagellum</u> must not be working.

11. (1 pt) It is missing a <u>food vacuole</u>, as that is where food is stored. <u>Gullet</u> is also acceptable.

12. (1 pt) It must have a <u>chloroplast</u> in which to store the chlorophyll.

13. (2 pts – one for each phylum) The multicellular algae are mostly found in phylum <u>Phaeophyta</u> and phylum <u>Rhodophyta</u>.

14. (3 pts – one for each letter. Give ½ point per letter if the student misses either the subkingdom or the phylum but gets the other correct.)

a. Note the hairy outside (cilia) and the two nuclei (macronucleus and micronucleus). Also, it is large compared to the others. This makes it part of <u>subkingdom Protozoa, phylum Ciliophora</u>.

b. Note the undefined shape. Those extensions are pseudopods. This makes it part of <u>subkingdom Protozoa, phylum Sarcodina</u>.

c. The fact that it is green is almost enough to place it in <u>subkingdom Algae, phylum Chlorophyta</u>. However, the spiral shapes should give it away as one of the organisms you studied as a part of that phylum.

Total possible points: 23

SOLUTIONS TO THE TEST FOR MODULE #4

1. (5 pts – one for each definition)

a. <u>Extracellular digestion</u> – Digestion that takes place outside of the cell

b. <u>Rhizoid hypha</u> – A hypha that is imbedded in the material on which the fungus grows

c. <u>Stolon</u> – An aerial hypha that asexually reproduces to make more filaments

d. <u>Fermentation</u> – The anaerobic breakdown of sugars into smaller molecules

e. <u>Hypha</u> – A filament of fungal cells

2. (1 pt) It is <u>parasitic</u>, because the haustorium is designed to invade the cells of the host and draw nutrients from them.

3. (1 pt) Since all multicellular fungi must have something that holds it to the material on which it grows, they all must have <u>rhizoid hyphae</u>. If the student answers with something relating to the mycelium, give him ½ of a point.

4. (1 pt – ½ for each) It provides <u>toughness</u> and <u>flexibility</u>.

5. (2 pts – one for each correct answer) The two most general answers are <u>sporophore</u> and <u>stolon</u>. However, the student can be more specific and include <u>sporangiophore</u> or <u>conidiophore</u> on the list. Please note that since these are specific kinds of sporophores, they cannot be counted if sporophore is given. Thus, an answer of "sporophore and conidiophore" would receive only one point, as a conidiophore is a specific kind of sporophore. The student could also include <u>an aerial hypha that absorbs oxygen</u>. We did not give that kind of aerial hypha a name, but it was discussed.

6. (4 pts – one for each letter)

a. Fungi that form spores on clublike basidia are in phylum <u>Basidiomycota.</u>
b. If we cannot determine the sexual mode of reproduction, we put the fungus in phylum <u>Deuteromycota</u>.
c. Fungi in phylum <u>Zygomycota</u> form zygospores right where the hyphae fuse.
d. If a fungus resembles both protozoa and fungi, it is in phylum <u>Myxomycota</u>.

7. (1 pt) The basidia are found on <u>the gills</u>. You can give ½ of a point for cap because it holds the gills.

8. (2 pts – one for the sexual mode and one for an asexual mode) <u>Sexually, bread molds reproduce when two mycelia form a zygospore</u>. Bread molds can <u>asexually reproduce when a stolon elongates and eventually starts another mycelium</u> or <u>when an aerial hypha forms a sporophore</u>.

9. (2 pts – ½ for each fungus and ½ for each malady) Any two of the following will work:

1) <u>rusts - crop damage</u>
2) <u>smuts - crop damage</u>
3) <u>ergot of rye (*Claviceps purpurea*) - death</u>
4) <u>*Cryphonectria parasitica* - chestnut blight</u>
5) <u>*Ophiostoma ulmi* - Dutch elm disease</u>
6) <u>*Synchytrium endobioticum* – potato wart</u>

10. (1 pt) <u>Most likely, the crop and the fungus form a mycorrhiza together</u>. The student can also describe the symbiotic relationship without giving the name "mycorrhiza."

11. (2 pts – one for the medicine, one for the general name) The medicine is called <u>penicillin</u>, and the general name for such a medicine is <u>antibiotic</u>.

12. (1 pt) It is probably in kingdom <u>Fungi</u>. It is most likely a yeast.

13. (1 pt) Slime molds appear in kingdom Protista in some biology books because their classification is in some dispute. <u>They resemble members of kingdom Protista during their feeding stage and members of kingdom Fungi during their reproductive stage</u>. Thus, they could belong to either kingdom.

Total possible points: 24

SOLUTIONS TO THE TEST FOR MODULE #5

1. (4 pts – one for each definition)

a. <u>Saturated fat</u> – A lipid made from fatty acids that have no double bonds between carbon atoms

b. <u>Physical change</u> – A change that affects the appearance but not the chemical makeup of a substance

c. <u>Model</u> – An explanation or representation of something that cannot be seen

d. <u>Isomers</u> – Two different molecules that have the same chemical formula

2. (1 pt) <u>An atom must have the same number of protons as electrons</u>. The student must have been reporting on an *ion*, which has different numbers of protons and electrons.

3. (1 pt) They <u>do not belong to the same element</u>. In order for them to belong to the same element, the number of protons must be the same.

4. (1 pt) The number after the element name tells you the total number of neutrons and protons in the nucleus. If there are 16 protons, there must be <u>18</u> neutrons for the total to add to 34.

5. (1 pt) There are 22 C's, 44 H's, and one O. That makes a total of <u>67</u> atoms.

6. (4 pts – one for each letter)

a. This is an <u>element</u>, as there is only one symbol and no number following the symbol.

b. This is a <u>molecule</u>, since two elements (two N's) are present.

c. This is a specific <u>atom</u>, since the name of the element and the total number of neutrons and protons in the nucleus are both given.

d. This is a <u>molecule</u>, as it is composed of two elements (P and H).

7. (1 pt) You are causing a <u>physical change</u>. All phase changes are physical changes.

8. (1 pt) This is an example of <u>diffusion</u>. Had it been osmosis, the water levels would have changed reflecting the motion of the solvent across the semipermeable membrane.

9. (2 pts – one for answering it is a product, and one for the number of molecules) It is a <u>product</u> because it is to the right of the arrow. There are <u>three</u> molecules produced because there is a "3" next to the chemical formula.

10. (2 pts – ½ for each answer) Photosynthesis requires <u>carbon dioxide</u>, <u>water</u>, <u>energy from sunlight</u>, and a <u>catalyst</u>

11. (1 pt) Remember, a carbohydrate has the same ratio of H's to O's as water (2:1). Thus, if there are two O's, there must be <u>four</u> H's.

12. (1 pt) Hydrolysis reactions break down big molecules like disaccharides into their constituent components, which are <u>monosaccharides</u>.

13. (1 pt – take off ½ for each wrong answer included) The pH scale says that pH's below 7 are acidic. The lower the pH, the more acidic. Thus, <u>A</u> is the only acidic one.

14. (1 pt – take off ½ for each wrong answer included) Acids must have an acid group. Thus, only A and C are acids. Unsaturated means there is at least one double bond between carbons. This tells us that <u>C</u> is the unsaturated fatty acid.

15. (1 pt) The properties of a protein are determined by the <u>type, number, and order of amino acids linked together</u>.

16. (1 pt) <u>The nucleotide's base</u> is what determines the "Morse code" of DNA.

17. (1 pt) <u>Hydrogen bonding</u> between the bases holds DNA's double helix together.

Total possible points: 25

SOLUTIONS TO THE TEST FOR MODULE #6

1. (6 pts – one for each definition)

a. Cytoplasmic streaming – The motion of cytoplasm in a cell that results in a coordinated movement of the cell's contents

b. Isotonic solution – A solution in which the concentration of solutes is essentially equal to that of the cell which resides in the solution

c. Cytolysis – The rupturing of a cell due to excess internal pressure

d. Phagocytosis – The process by which a cell engulfs foreign substances or other cells

e. Activation energy – Energy necessary to get a chemical reaction going

f. Plasmolysis – Collapse of a walled cell's cytoplasm due to a lack of water

2. (1 pt) It has performed secretion, because the protein will be used by other cells.

3. (1 pt) The middle lamella lies between the cell walls of plant cells.

4. (1 pt) Digestion breaks down big molecules for the purpose of respiration and biosynthesis, while respiration breaks down small molecules for the purpose of producing energy.

5. (2 pts – one for each answer) There are six correct answers and two acceptable ones. The student needs only identify two of the following: ribosomes, chloroplasts, smooth endoplasmic reticulum, rough endoplasmic reticulum, plastids, and Golgi bodies all have direct roles in biosynthesis. You can also count cytoplasm and nucleus, since they are involved in all cellular functions, including biosynthesis.

6. (1 pt) Rough ER has ribosomes, whereas smooth ER does not.

7. (1 pt) A chloroplast stores the pigment chlorophyll, and any organelle that stores pigments is generally called a chromoplast.

8. (1 pt) It must be sent to the lysosome so that it can be broken down into monosaccharides.

9. (1pt) It can still perform some respiration, but it will only do glycolysis, as that occurs in the cytoplasm.

10. (1 pt) Respiration in aerobic conditions provides more energy per molecule of glucose. In aerobic conditions, the cell can make 36 ATPs per molecule of glucose. In anaerobic conditions, it can make only two ATPs per molecule of glucose.

11. (1 pt) The electron transport system produces the most energy. It makes 32 ATPs, while the other two energy-producing steps each make only two ATPs.

12. (1 pt) <u>The fact that phospholipids have a hydrophilic end and a hydrophobic end</u> allows the plasma membrane to self-reassemble. The student need not use the terms "hydrophilic" and "hydrophobic." If the student mentions that it is the nature of the phospholipids that make up the plasma membrane, that's fine.

13. (1 pt – ½ for each stage) Both <u>glycolysis</u> and <u>the Krebs cycle</u> produce two ATPs.

14. (4 pts – one for each letter)

a. <u>protein</u> (The student could say active transport site)
b. <u>carbohydrate</u>
c. <u>cholesterol</u>
d. <u>filaments of the cytoskeleton</u>

Total possible points: 23

SOLUTIONS TO THE TEST FOR MODULE #7

1. (5 pts – one for each definition)

a. <u>Interphase</u> – The time interval between cellular reproduction

b. <u>Karyotype</u> – The figure produced when the chromosomes of a species during metaphase are arranged according to their homologous pairs

c. <u>Diploid cell</u> – A cell with chromosomes that come in homologous pairs

d. <u>Gametes</u> – Haploid cells (n) produced by diploid cells (2n) for the purpose of sexual reproduction

e. <u>Virus</u> – A non-cellular infectious agent that has two characteristics:
      (1) It has genetic material (RNA or DNA) inside a protective protein coat.
      (2) It cannot reproduce on its own.

2. (1 pt – ½ for each answer) <u>Environmental factors</u> and <u>spiritual factors</u> also affect a person's traits.

3. (5 pts – one for each letter and one for the proper order)

a. <u>anaphase</u>
b. <u>prophase</u>
c. <u>telophase</u>
d. <u>metaphase</u>

<u>The proper order is prophase, metaphase, anaphase, telophase.</u>

4. (2 pts – one for each answer) <u>The diploid number is 24</u>, since twelve pairs make a total of 24 chromosomes. <u>The haploid number is 12</u>, because the haploid number is just the number of pairs.

5. (2 pts – one for each answer) In mitosis, one cell becomes two cells that are identical to the original. Thus, if three cells undergo mitosis, <u>six cells result</u>. Since they are identical to the original cell, they will each have <u>32 chromosomes total</u>

6. (2 pts – one for each answer) In meiosis, one diploid cell becomes 4 haploid cells. Thus <u>12 cells result,</u> and they will have half of the original chromosome number. As a result, they will have <u>16 total chromosomes each</u>.

7. (1 pt) <u>Meiosis II</u> is essentially mitosis performed on haploid cells.

8. (1 pt) <u>Meiosis II</u> starts with a haploid cell and pulls apart the duplicated chromosomes.

9. (1 pt) <u>Mitosis</u> starts with diploid cells and ends with diploid cells.

10. (1 pt) Three of the four gametes are useless in <u>female</u> meiosis.

11. (1 pt) <u>It stimulates the body to produce antibodies against the virus.</u>

12. (1 pt – ½ for each answer) <u>DNA</u> and <u>RNA</u>

13. (2 pts – $\frac{2}{3}$ for each answer) Because uracil and adenine link together, and because cytosine and guanine link together, the codon must be: <u>uracil, adenine, cytosine</u>

14. (1 pt) It codes for <u>three</u> amino acids, because there are nine bases, and it takes three bases to code for each amino acid.

15. (3 pts - $\frac{1}{3}$ for each answer) Remember, DNA has thymine, not uracil. Thus, when there is a adenine on mRNA, there must have been a thymine on DNA. The DNA sequence must be: <u>adenine, cytosine, guanine, guanine, cytosine, thymine, adenine, thymine, thymine</u>

Total possible points: 29

SOLUTIONS TO THE TEST FOR MODULE #8

1. (6 pts – one for each definition)

a. <u>True breeding</u> – If an organism has a certain characteristic that is always passed on to its offspring, we say that this organism bred true with respect to that characteristic.

b. <u>Allele</u> – One of a pair of genes that occupies the same position on homologous chromosomes

c. <u>Recessive allele</u> – An allele that will not determine the phenotype unless the genotype is homozygous in that allele

d. <u>Monohybrid cross</u> – A cross between two individuals, concentrating on only one definable trait

e. <u>Dihybrid cross</u> – A cross between two individuals, concentrating on two definable traits

f. <u>Autosomal inheritance</u> – Inheritance of a genetic trait not on a sex chromosome

2. (4 pts – one for each principle)

   1. <u>The traits of an organism are determined by its genes.</u>

   2. <u>Each organism has two alleles that make up the genotype for a given trait.</u>

   3. <u>In sexual reproduction, each parent contributes ONLY ONE of its alleles to its offspring.</u>

   4. <u>In each genotype, there is a dominant allele. If it exists in an organism, the phenotype is determined by that allele.</u>

3. (1 pt – ½ for each answer) To roll his tongue, a person needs only one dominant allele. Thus, the person could have genotypes of <u>RR</u> or <u>Rr</u>.

4. (1 pt) A gamete has <u>one</u> allele for each trait. That way, when two gametes fertilize each other, the zygote will have two alleles for each trait.

5. (1 pt) Normal cells have <u>two</u> alleles for each trait.

6. a. (1 pt) Heterozygous means one of each allele. Thus, his genotype must be <u>Tt</u>.

   b. (1 pt) Not being able to taste PTC is recessive. To express that trait, she must be <u>tt</u>.

   c. (1 pt – ½ for each genotype) The Punnett square for this situation is given on the next page.

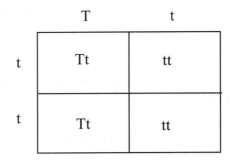

This means 50% will be Tt and 50% will be tt.

7. a. (2 pts – one for the right column and row labels, and ¼ for each box) Since the man is not hemophilic, he must have the dominant trait as his only allele. Since the woman does not have the disease but does carry it, she must have one of each allele. The Punnett square, then, looks like this:

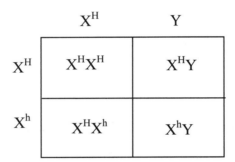

Whether $X^H Y$ is on top or on the left is not important.

b. (1 pt) Boxes with two X's represent girls. Since none of those have both recessive alleles, 0% of the girls will have hemophilia.

c. (1 pt) Boxes with an "X" and a "Y" represent boys. Since half of those boxes have only the recessive allele, 50% of the boys will have hemophilia.

8. (1 pt) In sex-linked traits, there is no allele on the Y chromosome. Thus, males have only one allele. Your chance of expressing the recessive phenotype is better if you have only one allele.

9. (1 pt) Genetics is only one part of a person's makeup. Environmental and spiritual factors play a role as well.

10. (2 pts – ½ for each) The student needs four of the following: Autosomal inheritance, sex-linked inheritance, change in chromosome structure, allele mutation, or change in chromosome number.

11. a. (1 pt) You need all possible combinations of the four alleles when one allele for color combines with one allele for height. The possibilities are *YT, Yt, yT, yt*. The order is irrelevant.

b. (2 pts – ½ for the row labels, ½ for the column labels, and one for all of the boxes) The Punnett square shown on the next page:

	YT	Yt	yT	yt
YT	YYTT	YYTt	YyTT	YyTt
Yt	YYTt	YYtt	YyTt	Yytt
yT	YyTT	YyTt	yyTT	yyTt
yt	YyTt	Yytt	yyTt	yytt

c. (2 pts – ½ for each phenotype. If the student lists genotypes and not phenotypes, give 1 point.)

tall, yellow peas (genotypes YYTT, YyTt, YyTT, YYTt) 9 of 16 or 56.25 %
tall, green peas (genotypes yyTT, yyTt) 3 of 16 or 18.75 %
short, yellow peas (genotypes YYtt, Yytt) 3 of 16 or 18.75 %
short, green peas (genotype yytt) 1 of 16 or 6.25 %

12. (1 pt) Filled circles and squares represent those who cannot roll their tongues. Remember, if the same trait is expressed in both parents and they give rise to the other trait, they must each carry a recessive allele. Look at the first set of parents. They are both represented by empty shapes. Yet, they give rise to a child with a filled square. Thus, the empty shapes must represent the dominant allele and the filled must represent the recessive.

Total possible points: 30

SOLUTIONS TO THE TEST FOR MODULE #9

1. (6 pts – one for each definition)

a. The immutability of species – The idea that each individual species on the planet was specially created by God and could never fundamentally change

b. Microevolution – The theory that natural selection can, over time, take an organism and transform it into a more specialized species of that organism

c. Macroevolution – The hypothesis that processes similar to those at work in microevolution can, over eons of time, transform an organism into a completely different kind of organism

d. Strata – Distinct layers of rock

e. Fossils – Preserved remains of once-living organisms

f. Structural Homology – The study of similar structures in different species

2. (1 pt) He traveled on The *HMS Beagle*.

3. (2 pts – one for each idea) The concept of a struggle for survival as proposed by Malthus and the present is the key to the past as proposed by Lyell both influenced Darwin greatly.

4. (1 pt) This is an example of microevolution. The horses remained horses, they just got faster.

5. (1 pt) This would be macroevolution, as the bacteria evolved into a completely different organism.

6. (1 pt) None of the data sets discussed in this module provide conclusive evidence for macroevolution.

7. (3 pts – one for each data set) Structural homology, the fossil record, and molecular biology all provide conclusive evidence against macroevolution.

8. (1 pt) Darwin proposed the theory of microevolution in that book. It is a well-documented scientific theory today. The wrong ideas in the book are those related to macroevolution.

9. (1 pt) The ape's should be closer to the human's, because according to the hypothesis of macroevolution, the ape is closer in lineage to the man than is the rat.

10. (1 pt) Neo-Darwinism uses mutations to add information to the genetic code.

11. (1 pt) Punctuated equilibrium attempts to explain why there are no intermediate links in the fossil record.

12. (1 pt) Mutation is not the only way a bacterium becomes immune to an antibiotic. Conjugation and transformation are the *main* means by which bacteria become immune to an antibiotic. Transduction can also do this.

13. (1 pt) The fact that representatives of all major animal phyla can be found in some of the lowest sedimentary rock in the geological column is often referred to as the <u>Cambrian explosion</u>.

14. (2 pts) <u>The sequence in (a) is closest to the sequence of interest</u>.

Total possible points: 23

SOLUTIONS TO THE TEST FOR MODULE #10

1. (5 pts – one for each definition)

a. <u>Ecosystem</u> – An association of living organisms and their physical environment

b. <u>Biomass</u> – A measure of the total dry mass of organisms within a particular region

c. <u>Watershed</u> – An ecosystem where all water runoff drains into a single body of water

d. <u>Transpiration</u> – Evaporation of water from the leaves of a plant

e. <u>Greenhouse effect</u> – The process by which certain gases (principally water vapor, carbon dioxide, and methane) trap heat that would otherwise escape the earth and radiate into space

2. (2 pts – one for recognizing that the ecosystem would be thrown out of balance, one for specifically mentioning the frog) <u>The frog population would grow out of control because frogs would have no predators</u>. Even though the praying mantis, plants, mouse, and grasshopper would lose predators, they would still have other predators (the song bird eats the praying mantis and grasshopper, for example). Thus, their populations would not grow unchecked. The frog would lose all of its predators, however, which would cause the frog population to grow unchecked.

3. (2 pts – one for each level) When the owl eats the rabbit, squirrel, or mouse, it is a <u>secondary consumer</u>. When it eats the snake, it is a <u>tertiary consumer</u>.

4. (2 pts – one for each level) When it eats plants, it is a <u>primary consumer</u>. When it eats the praying mantis or the frog, it is a <u>tertiary consumer</u>.

5. (1 pt) The key here is that each level should be roughly half as wide as the one under it.

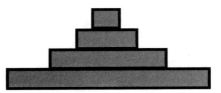

6. (4 pts – one for each relationship and one for each description) The student need give only two of these three:

<u>The clownfish and the sea anemone</u> form a mutualistic relationship. <u>The clownfish is protected by the sea anemone, and it attracts food to the sea anemone.</u>

<u>The goby and the blind shrimp</u> have a mutualistic relationship in which <u>the goby protects the blind shrimp and the blind shrimp provides a home for the goby.</u>

<u>The Oriental sweetlips and blue-streak wrasse</u> form a mutualistic relationship in which <u>the sweetlips gets it teeth cleaned by the wrasse and the wrasse gets food from the sweetlips' teeth</u>.

7. (1 pt) Too many nutrients in the water indicate that there are no plants to regulate the nutrient flow in the water cycle. <u>The most likely explanation is a lack of plant life in the watershed</u>. This could happen through deforestation, for example.

8. (1 pt) <u>Photosynthesis adds oxygen to the air</u>.

9. (1 pt) <u>Carbon dioxide can be dissolved in the ocean</u>.

10. (1 pt) Without surface runoff, more water would evaporate from the ocean than what gets returned to it. <u>The ocean would begin losing water</u>.

11. (1 pt) <u>The greenhouse effect is a good thing because without it, the earth would be too cold for life to survive. Global warming is an enhancement of the greenhouse effect that causes the earth to get too warm</u>. Thus, it really is too much of a good thing.

12. (1 pt) The most reliable data indicate that <u>global warming is *not* happening today</u>.

13. (2 pts – one for nitrogen and one for producers) Nitrogen-fixing bacteria provide <u>nitrogen</u> that <u>producers</u> can use. Consumers get their nitrogen from the producers.

Total possible points: 24

SOLUTIONS TO THE TEST FOR MODULE #11

1. (6 pts – one for each definition)

a. <u>Invertebrates</u> – Animals that lack a backbone

b. <u>Vertebrates</u> – Animals that possess a backbone

c. <u>Nematocysts</u> – Small capsules that contain a toxin which is injected into prey or predators

d. <u>Posterior end</u> – The end of an animal that contains its tail

e. <u>Hermaphroditic</u> – Possessing both the male and the female reproductive organs

f. <u>Mantle</u> – A sheath of tissue that encloses the vital organs of a mollusk, makes the mollusk's shell, and performs respiration

2. (5 pts – one for each letter)

a. <u>Porifera</u>
b. <u>Mollusca</u>
c. <u>Cnidaria</u>
d. <u>Annelida</u>
e. <u>Platyhelminthes</u>

3. (1 pt) These are <u>bivalves</u>, because they have two shells.

4. (2 pts – one for what and one for how) <u>Sponges eat the algae, bacteria, and organic matter that is in the water.</u> The student need only list one of the food types. They get this food by <u>pumping water through their bodies and capturing the food from the water.</u>

5. (1 pt) <u>Hard, prickly sponges contain spicules while soft sponges contain spongin.</u>

6. (1 pt) <u>The organisms of phylum Cnidaria have nematocysts.</u>

7. (1 pt) <u>It has not mated yet.</u> When an earthworm mates, its seminal vesicles are emptied and its seminal receptacles get filled.

8. (1 pt) <u>The cuticle allows the earthworm to get oxygen and release carbon dioxide.</u> The student could just say that it performs respiration.

9. (1 pt) <u>The second planarian is the parasite.</u> Parasites can have simple nervous systems because they need not search for food.

10. (7 pts – one for each letter)

a. <u>ganglia</u> b. <u>esophagus</u> c. <u>oviduct</u> d. <u>dorsal blood vessel</u> e. <u>clitellum</u> f. <u>seminal vesicle</u>
g. <u>ventral nerve cord</u>

11. (1 pt) Members of phylum Cnidaria asexually reproduce by <u>budding</u>.

12. (1 pt) Members of phylum Platyhelminthes asexually reproduce by <u>regeneration</u>.

Total possible points: 28

SOLUTIONS TO THE TEST FOR MODULE #12

1. (5 pts – one for each definition)

a. <u>Exoskeleton</u> – A body covering, typically made of chitin, that provides support and protection

b. <u>Simple eye</u> – An eye with only one lens

c. <u>Open circulatory system</u> – A circulatory system that allows the blood to flow out of the blood vessels and into various body cavities so that the cells are in direct contact with the blood

d. <u>Statocyst</u> – The organ of balance in a crustacean

e. <u>Gonad</u> – A general term for the organ that produces gametes

2. (3 pts – one for each letter) a. <u>antennae</u> b. <u>carapace</u> c. <u>swimmerets</u>

3. (7 pts – one for each letter) a. <u>stomach</u> b. <u>heart</u> c. <u>pericardial sinus</u> d. <u>anus</u> e. <u>nerve cord</u> f. <u>digestive glands</u> g. <u>green gland</u>

4. (1 pt) <u>Arthropods molt because their growing bodies get too large for their exoskeletons.</u>

5. (2½ pts – ½ for each characteristic) The common characteristics of arachnids are <u>four pairs of walking legs</u>, <u>two segments in body</u>, <u>no antennae</u>, <u>book lungs</u>, and <u>four pairs of simple eyes</u>.

6. (2 pts – ½ for each characteristic) The common characteristics of insects are <u>three pairs of walking (or jumping) legs</u>, <u>wings</u>, <u>three segments in the body</u>, and <u>one pair of antennae</u>.

7. (1½ pts – ½ for each stage) The stages of incomplete metamorphosis are <u>egg</u>, <u>nymph</u>, and <u>adult</u>.

8. (1 pt) <u>A complex network of tracheas</u> connected to spiracles in the exoskeleton takes the place of a respiratory system in insects.

9. (1 pt) Horny wings indicate that it is a beetle from order <u>Coleoptera</u>.

10. (1pt) Leathery wings protecting membranous wings are typically found in order <u>Orthoptera</u>.

Total possible points: 25

SOLUTIONS TO THE TEST FOR MODULE #13

1. (3 pts – one for each definition)

a. <u>Bone marrow</u> – A soft tissue inside the bone that produces blood cells

b. <u>Cerebrum</u> – The lobes of the brain that integrate sensory information and coordinate the creature's response to that information

c. <u>Ectothermic</u> – Lacking an internal mechanism for regulating body heat

2. (2 pts – one for each organism) There are many possible answers here. <u>Lampreys, lancelets, sea squirts, frogs, toads, and salamanders</u> all have larval stages. The student needs list only two.

3. (7 pts – one for each letter)

a. <u>Class Chondrichthyes</u>   b. <u>Class Osteichthyes</u>   c. <u>Subphylum Urochordata</u>   d. <u>Class Amphibia</u>
e. <u>Class Amphibia</u>   f. <u>Subphylum Cephalochordata</u>   g. <u>Class Agnatha</u>

4. (1 pt) <u>Sight is its weakest sense</u>, because the optic lobes control sight.

5. (1 pt) <u>Hemoglobin</u> carries oxygen in red blood cells.

6. (2 pts – one for each answer) Fertilization is <u>external</u> and the development is <u>oviparous</u>.

7. (1 pt) The shark has a cartilaginous skeleton. The other two have bony skeletons. Thus, <u>the shark has the most flexible skeleton.</u>

8. (4 pts – one for each letter) a. <u>Spinal cord</u>   b. <u>Air bladder</u>   c. <u>Gonad</u>   d. <u>Liver</u>

9. (4 points – one for each function)

Organ	Basic Function
Spinal cord	Sends messages from brain to other parts of the body and vice-versa
Air bladder	Allows fish to change depths and float in water
Gonad	Reproduction
Liver	Makes bile for the digestion of fats and does many other chemical tasks

10. (3 pts – one for each letter) a. <u>artery</u>   b. <u>vein</u>   c. <u>artery</u>

11. (3 pts – one for each letter) a. <u>oxygen rich</u>   b. <u>oxygen poor</u>   c. <u>oxygen poor</u>

12. (1 pt) An amphibian's <u>skin</u> is its most important respiratory organ.

Total possible points: 32

SOLUTIONS TO THE TEST FOR MODULE #14

1. (5 pts – one for each definition)

a. <u>Vegetative organs</u> – The parts of a plant (such as stems, roots, and leaves) that are not involved in reproduction

b. <u>Reproductive plant organs</u> – The parts of a plant (such as flowers, fruits, and seeds) involved in reproduction

c. <u>Undifferentiated cells</u> – Cells that have not specialized in any particular function

d. <u>Phloem</u> – Living vascular tissue that carries sugar and organic substances throughout a plant

e. <u>Deciduous plant</u> – A plant that loses its leaves for winter

2. (6 pts – ½ for each shape, margin, and venation)

Letter	Shape	Margin	Venation
a.	Cordate	Serrate	Palmate
b.	Deltoid	Dentate	Palmate
c.	Linear	Entire	Parallel
d.	Elliptical	Undulate	Pinnate

3. (1 pt) Guard cells <u>open and close the stomata.</u>

4. (1 pt) <u>The top layer will be lighter.</u> The side of the leaf with the spongy mesophyll is lighter than the side with the palisade mesophyll.

5. (1 pt) <u>The abscission layer</u> controls when the leaves die and fall off a deciduous plant.

6. (2 pts – one for what they are and one for what they do) Anthocyanins are <u>a group of pigments that can give a leaf a color other than green</u>.

7. (1 pt) Undifferentiated cells are found in the <u>meristematic tissue</u>.

8. (1 pt) <u>It is from a monocot.</u> The face-like appearance indicates this.

9. (1 pt) <u>Bark is cracked because the stem grows and breaks the bark when it gets too big.</u>

10. (1 pt) Trees that produce cones are members of <u>phylum Coniferophyta</u>.

11. (1 pt) <u>This plant has a taproot system.</u>

12. (2 pts – one for each difference) There are several differences. The student need list only two. <u>The number of cotyledons produced in the seed (one in monocots, two in dicots)</u>, <u>the venation of the leaves (parallel in monocots, netted in dicots)</u>, <u>the fibrovascular bundles (different appearances and packaging in the stems and roots)</u>, <u>and the number of petals on their flowers (groups of three or six in monocots and groups of four or five in dicots</u>.

Total possible points: 23

SOLUTIONS TO THE TEST FOR MODULE #15

1. (6 pts – one for each definition)

a. <u>Physiology</u> – The study of life processes in an organism

b. <u>Nastic movement</u> – A plant's response to a stimulus such that the direction of the response is preprogrammed and not dependent on the direction of the stimulus

c. <u>Pore spaces</u> – Spaces in the soil that determine how much water and air the soil can hold

d. <u>Translocation</u> – The process by which organic substances move through the phloem of a plant

e. <u>Gravitropism</u> – A growth response to gravity

f. <u>Imperfect flowers</u> – Flowers with either stamens or carpels, but not both

2. (2 pts – ½ for each process) A plant uses water for <u>photosynthesis</u>, <u>turgor pressure</u>, <u>hydrolysis</u>, and <u>transport</u>.

3. (1 pt) <u>Substances will not flow through the xylem</u>. The flow of water and minerals through the xylem is most likely caused by the evaporation of water through the stomata, as dictated by the cohesion-tension theory.

4. (1 pt) <u>Substances will flow through the phloem</u>. The flow of materials through the phloem is controlled by phloem cells and is unrelated to the position of the stomata. The flow will *eventually* stop, because the plant will starve due to a lack of photosynthesis. That will take a long while, however, because plants store up excess food.

5. (1 pt) <u>It came from the phloem</u>. The xylem transport mostly water and minerals, whereas the phloem transport organic substances.

6. (1 pt) <u>It will not starve to death</u>. Insectivorous plants do not use the insects they catch for food. They use them for the raw materials of biosynthesis.

7. (1 pt) <u>The parent reproduced vegetatively</u>. Vegetative reproduction is asexual, which results in a genetic copy.

8. (4 pts – one for each letter) a. <u>Ovules</u>   b. <u>Anther</u>   c. <u>Sepal</u>   d. <u>Petal</u>

9. (1 pt) <u>b</u>

10. (1 pt) <u>c</u>

11. (1 pt) <u>d</u>

12. (1 pt) <u>a</u>

13. (1 pt) <u>They feed the embryo</u> by either being the food source or transferring food from the endosperm to the embryo. The student needs only mention that cotyledons feed the embryo.

14. (1 pt) A fruit <u>allows for the dispersal of seeds away from the parent</u>.

15. (3 pts) There are many possible answers. The student needs list only three:

<u>wind, bees, beetles, birds, moths, or butterflies</u>

16. (1 pt) <u>An embryo sac has more cells.</u> A pollen grain usually has two or three: one or two sperm cells and a tube nucleus. An embryo sac has seven: six haploid eggs and a double-nucleus cell that becomes the endosperm.

17. (1 pt) The zygote is diploid, because it is the result of a union between two haploid cells. The <u>endosperm</u> is the result of a union between three haploid nuclei, so it is a "3n" cell.

Total possible points: 28

SOLUTIONS TO THE TEST FOR MODULE #16

1. (5 pts – one for each definition)

a. <u>Amniotic egg</u> – A shelled, water-retaining egg that allows reptile, bird, and certain mammal embryos to develop on land

b. <u>Hemotoxin</u> – A poison that attacks the red blood cells and blood vessels, destroying circulation

c. <u>Endotherm</u> – An organism that is internally warmed by a heat-generating metabolic process

d. <u>Placenta</u> – A structure that allows an embryo to be nourished with the mother's blood supply

e. <u>Gestation</u> – The period of time during which an embryo develops before being born

2. (1 pt) <u>It is a bird</u>. Not all bird characteristics are listed, but no vertebrates other than birds have all of the characteristics listed in this question.

3. (1 pt) <u>It is a reptile</u>. Not all reptile characteristics are listed, but no vertebrates other than reptiles have all of the characteristics listed in this question.

4. (1 pt) <u>It is a mammal</u>. Not all mammal characteristics are listed, but no vertebrates other than mammals have all of the characteristics listed in this question.

5. (4 pts – one for each letter) a. <u>Rhynchocephalia</u>   b. <u>Squamata</u>   c. <u>Crocodilia</u>   d. <u>Testudines</u>

6. (1 pt) <u>It is not possible</u>. Living reptiles are ectothermic. Some scientists speculate that some extinct reptiles might have been endothermic.

7. (1 pt) <u>Contour feathers</u> have hooked barbules.

8. (1 pt) You have found a <u>reptile egg</u>. Bird eggs are covered with a hard shell.

9. (1 pt) <u>The bird should start preening</u>. This will oil the feathers, making the hooked barbules slide easily on the smooth barbules.

10. (1 pt) <u>The underhair will be thicker</u>, because the main function of underhair is insulation.

11. (1 pt) <u>The first species has the shorter gestation period</u>. The longer the gestation period, the more developed the offspring are at birth. Since the second species gave birth to well-developed offspring, it must have had a long gestation period.

12. (4 pts – one for each letter) a. <u>amnion</u>   b. <u>allantois</u>   c. <u>yolk sac</u>   d. <u>albumen</u>

Total possible points: 22

QUARTERLY TEST #1

1. What are the four criteria for life?

2. An organism is classified as an omnivore. Is it a heterotroph or an autotroph? Is it a producer, consumer, or decomposer?

3. A scientist observes a process and then comes up with an idea that he or she thinks explains that process. At that point, is the idea a hypothesis, theory, or scientific law?

4. Name the classification groups in our hierarchical classification scheme in order.

5. An organism is a multicellular decomposer made of eukaryotic cells. To what kingdom does it belong?

6. An organism is a single-celled consumer made of prokaryotic cells. To what kingdom does it belong?

7. What is the difference between an aerobic process and an anaerobic process?

8. What do most bacteria use for locomotion?

9. What is the main mode of reproduction in bacteria?

10. Briefly describe conjugation in bacteria.

11. Even though conjugation among bacteria does not result in offspring, it can significantly affect the population of a bacteria growth. Why?

12. What are the technical names of the three common bacterial shapes?

13. What is the function of a contractile vacuole? What is the difference between this and a food vacuole?

14. Name at least two pathogenic organisms from kingdom Protista.

15. For each of the phyla below, list the means of locomotion employed by the organisms in that phylum:

Sarcodina, Mastigophora, Ciliophora

16. What is mutualism?

17. Which phylum (Sarcodina, Mastigophora, Ciliophora, Sporozoa, Chlorophyta, Chrysophyta, Pyrrophyta, Phaeophyta, Rhodophyta) contains the organisms responsible for most of the photosynthesis that occurs on earth? What generic term is used to refer to these organisms?

18. What two phyla (see the list in the question above) contain mostly macroscopic algae?

19. Many biologists say that a mushroom is much like an iceberg, because only about 10% of an iceberg is visible from the surface of the ocean. What do they mean?

20. You observe the mycelium of a fungus under a microscope and notice no distinct cell walls but several cell nuclei. Is the mycelium composed of septate hyphae or nonseptate hyphae?

21. What is the function of each of the following specialized hyphae?

        rhizoid hyphae   stolon   sporophore   haustorium

22. What phylum contains fungi that do not have a known mode of sexual reproduction?

23. Name a pathogenic fungus and the malady that it causes.

24. When a slime mold reproduces, it resembles organisms from what kingdom?

25. Name a form of mutualism in which fungi participate. Describe the relationship and the job of each participant in that relationship.

QUARTERLY TEST #2

1. To turn a gas into a liquid, do you remove or add energy?

2. What determines the vast majority of characteristics in an atom?

3. How many electrons are in an atom that has 32 protons?

4. How many atoms (total) are in a molecule of $C_4H_8O_2$? What atoms are present and how many of each atom?

5. Two solutions of different solute concentration are separated by a membrane. After a while, the water levels of the two solutions change. Has osmosis or diffusion taken place? What kind of membrane is being used?

6. Consider the following chemical reaction:

$$O_2 + 2H_2 \rightarrow 2H_2O$$

   a. What are the reactants?
   b. What are the products?
   c. How many molecules of H_2 are used in the reaction?

7. Describe the pH scale and what it measures.

8. What are the basic building blocks of proteins?

9. How does DNA store information?

10. What are the four stages of cellular respiration? Which one produces the most energy?

11. If a cell has no oxygen, what stage(s) of cellular respiration can still run?

12. A DNA strand has the following sequence of nucleotides:

   guanine, cytosine, adenine, adenine, thymine, guanine

   a. What will the mRNA sequence be?

   b. How many amino acids will the mRNA code for?

13. List (in order) the four stages of mitosis.

14. The diploid number of a cell is 16. What is its haploid number?

15. What is the difference between a gamete and a normal cell?

16. List (in order) all of the stages of meiosis.

17. What are gametes produced in animal males called? What are gametes produced in animal females called?

18. If a virus uses DNA as its genetic material, is it alive? Why or why not?

19. A person decides to wait until he contracts measles before getting the vaccine. What is wrong with this strategy?

20. Three pea plants have the following alleles for tall ("T") and short ("t") heights. What is the phenotype of each? Note whether it is homozygous or heterozygous.

a. TT   b. Tt   c. tt

21. The allele for axial flowers ("A") in a pea plant is dominant over the allele for terminal flowers ("a"). A pea plant which is homozygous in the axial flower allele is crossed with a pea plant that is heterozygous. What are the possible genotypes and phenotypes, along with their percentage chances, for the offspring?

22. Give the possible phenotypes and the percentage chance for each in the dihybrid cross between two pea plants that are heterozygous in producing smooth, yellow peas. The smooth allele is dominate over the wrinkled allele, and yellow is dominant over green.

23. In fruit flies, the color of the eye is a genetic trait that is sex-linked. What is the percentage of males that will have white eyes when a heterozygous, red-eyed female is crossed with a white-eyed male? What is the percentage of females that will have white-eyes from the same cross?

24. Two individuals have the exact same genotype for a certain trait but they are not identical when it comes to that trait. How is this possible?

25. If a recessive genetic disease occurs much more frequently in men than women, which chromosome should be studied as a possible source of the disease?

QUARTERLY TEST #3

1. What change to the concept of macroevolution did Neo-Darwinism make?

2. What age-old concept was Darwin able to dispel with his research?

3. How would an adherent to punctuated equilibrium explain the lack of intermediate links in the fossil record?

4. Does the study of structural homology provide evidence for or against macroevolution? Why?

5. Did Darwin ever recant his scientific beliefs?

6. The amino acid sequences in the protein called cytochrome C are studied for many different organisms. The sequences are all compared to that of a horse. According to the macroevolution hypothesis, which should be more similar to the cytochrome C of a horse: the cytochrome C of a fish, the cytochrome C of a giraffe, or the cytochrome C of a bacterium?

7. If the data discussed in problem #6 were actually analyzed, would the result be as predicted by the macroevolution hypothesis?

8. Give two examples of mutualism in creation.

9. What is the principal means by which oxygen is taken from the air?

10. What is the principal means by which oxygen is restored to the air?

11. Consider the following ecological pyramid

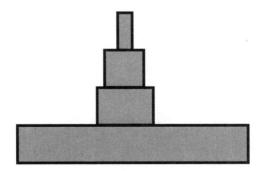

Between which two trophic levels is the smallest amount of energy wasted?

12. Is global warming occurring now?

13. Are the vast majority of animals vertebrates or invertebrates?

14. How do sponges get their prey?

15. What roles do amebocytes play in the life of a sponge?

16. Why do cnidarians not need respiratory or excretory systems?

17. If a jellyfish reproduces sexually, what form is it in?

18. What will happen to an earthworm if its cuticle gets dry?

19. If a flatworm has complex nervous and digestive systems, is it most likely free-living or parasitic?

20. Name the five common characteristics among the arthropods.

21. What happens when a crayfish loses a limb?

22. Why do arthropods molt?

23. What five characteristics set arachnids apart from the other arthropods?

24. What four characteristics set insects apart from the other arthropods?

25. Why don't insects have respiratory systems?

QUARTERLY TEST #4

1. What does the cerebrum do?

2. To which class does a lamprey belong?

3. What is the name of the cells that carry oxygen in the blood?

4. Which has the most flexible skeleton: a salmon, a carp, or a shark?

5. A female mates with a male and then lays eggs that develop and hatch. Is this external or internal fertilization? What kind of development is this?

6. What does a fish use to feel vibrations in the water?

7. What is the difference between a vein and an artery?

8. A botanist examines a portion of a plant and finds a lot of meristematic tissue. Is that portion of the plant still growing or has it stopped growing?

9. Determine the shape, margin, and venation of the following leaves:

Illustrations from the MasterClips collection

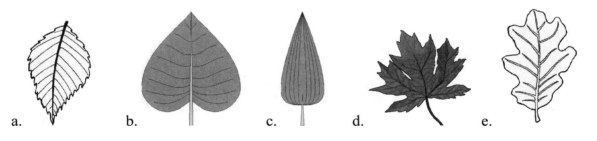

a.   b.   c.   d.   e.

10. Why is the bottom of a leaf typically a lighter shade of green than the top of the leaf?

11. Fill in the blank: A tree with an abscission layer is _____.

12. What phylum contains plants that must be small?

13. What is the fundamental difference between monocots and dicots?

14. What property of water governs how water flows upwards in a plant?

15. Vascular plants have two kinds of structures for the transport of water and nutrients. What are they called?

16. Of the two structures listed in Problem #15, which must be made of living cells in order to be able to do their job?

17. What is the male reproductive organ of a flower? What is the female reproductive organ?

18. What is the purpose of a cotyledon?

19. How do insects aid in the physiology of flowering plants?

20. State the five characteristics that set reptiles apart from other vertebrates.

21. Which order (Rhynchocephalia, Squamata, Crocodilia, Testudines) contains lizards?

22. State the six characteristics that set birds apart from other vertebrates.

23. What is a bird actually doing when it is preening?

24. State the five characteristics that set mammals apart from other vertebrates.

25. What kind of hair serves as insulation for a mammal?

SOLUTIONS TO QUARTERLY TEST #1

1. (4 pts – one for each criterion)

- All life forms contain deoxyribonucleic acid (DNA).

- All life forms have a method by which they extract energy from the surroundings and convert it into energy that sustains them.

- All life forms can sense changes in their surroundings and respond to those changes.

- All life forms reproduce.

2. (2 pts – one for each answer) Omnivores eat plants and nonplants. This means they depend on other organisms for food, making them heterotrophs, which are also known as consumers.

3. (1 pt) The idea is a hypothesis. Until it is tested with a lot of data, it cannot become a theory.

4. (2 pts – one for the correct names, one for the correct order) Kingdom, Phylum, Class, Order, Family, Genus, Species

5. (1 pt) It belongs in kingdom Fungi. Since it is multicellular, it is not in Monera or Protista. In addition, it is not in Plantae because it is not an autotroph, and it is not in Animalia because animals are consumers. Kingdom Fungi contains the majority of decomposers in creation.

6. (1 pt) It is in kingdom Monera. Prokaryotic cells belong to this kingdom.

7. (1 pt) Aerobic processes use oxygen, while anaerobic processes do not.

8. (1 pt) Most bacteria use flagella for locomotion.

9. (1 pt) Asexual reproduction is the main form of reproduction among bacteria.

10. (1 pt) Conjugation passes a trait from one bacterium to another via a small strand of DNA called a plasmid. If the students goes into the steps involved in conjugation, that's fine. However, mentioning the transfer of the plasmid is all that is necessary.

11. (1 pt) If the trait that is passed by the plasmid allows bacteria to survive when they otherwise wouldn't, it will increase the population.

12. (3 pts – one for each name) The shapes are coccus (spherical), bacillus (rod shaped), and spirillum (helical). The student must use the terms underlined, as those are the technical names.

13. (2 pts – one for the function of the contractile vacuole and one for the function of the food vacuole) A contractile vacuole collects excess water in a cell and expels it to reduce the pressure inside the cell. This keeps the cell from exploding. The food vacuole, on the other hand, stores food while it is being digested and has nothing to do with excess water or pressure.

14. (2 pts – one for each organism) There are more than two, but the student need list only two. The ones we discussed in Module #3 are *Entamoeba histolytica*, *Trypanosoma*, *Balantidium coli*, *Plasmodium*, and *Toxoplasma*.

15. (3 pts – one for each mode of locomotion) Sarcodina: pseudopods, Mastigophora: flagella, Ciliophora: cilia.

16. (1 pt) In mutualism, two or more organisms live together in a mutually beneficial way. In many cases of mutualism, it is impossible for one organism to live without the other.

17. (2 pts – one for the phylum and one for the generic name) Phylum Chrysophyta contains the diatoms, which are responsible for most of the world's photosynthesis.

18. (2 pts – one for each phylum) Phylum Phaeophyta and phylum Rhodophyta contain mostly macroscopic algae.

19. (1 pt) Typically, we see only the fruiting body of a mushroom. Like an iceberg, that visible part is only a small fraction of the total mushroom, because the mycelium is the largest component of a mushroom.

20. (1 pt) The fungus is composed of nonseptate hyphae, because nonseptate hyphae do not have distinct cell walls but have distinct nuclei.

21. (4 pts – one for each function) Rhizoid hyphae support the fungus and digest the food; a stolon asexually reproduces; a sporophore releases spores for reproduction; and a haustorium invades the cells of a living host to absorb food directly from the cytoplasm.

22. (1 pt) Phylum Deuteromycota contains the imperfect fungi, which have no known mode of sexual reproduction.

23. (1 pt) There are many pathogenic fungi. The student need only list one:

1) rusts - crop damage   4) *Cryphonectria parasitica* - chestnut blight
2) smuts - crop damage   5) *Ophiostoma ulmi* - Dutch elm disease
3) ergot of rye (*Claviceps purpurea*) - death 6) *Synchytrium endobioticum* – potato wart

24. (1 pt) When a slime mold reproduces, it resembles organisms from kingdom Fungi.

25. (2 pts – one for the relationship and one for the description) Fungi participate in mutualism by forming lichens and mycorrhizae. A lichen is a mutualistic relationship between a fungus and an alga. The alga produces food for both creatures via photosynthesis, and the fungus supports and protects the alga. Mycorrhizae are mutualistic relationships between a fungus's mycelium and a plant's root system. The mycelium takes nutrients from the root, while it collects minerals from the soil and gives it to the root. The student need list only one of these relationships.

Total possible points: 42

SOLUTIONS TO QUARTERLY TEST #2

1. (1 pt) You <u>remove energy</u> to turn a gas into a liquid.

2. (1 pt) <u>The number of electrons (or protons) in an atom determines the vast majority of its characteristics</u>.

3. (1 pt) Since atoms have the same number of electrons and protons, there must be <u>32 electrons</u>.

4. (2 pts – ½ for each answer) The subscripts after the elemental abbreviations tell you how many of each atom is in the molecule. Thus, there are <u>four carbons</u>, <u>eight hydrogens</u>, and <u>two oxygens</u>, for a <u>grand total of 14 atoms</u>.

5. (2 pts – one for each answer) Since the water levels changed, that means solvent traveled from one side of the membrane to the other, but solute did not. This is <u>osmosis</u>, which requires a <u>semipermeable membrane</u>.

6. a. (1 pt – ½ for each reactant) Reactants appear to the left of the arrow. The numbers to the left of the chemical formulas, however, do not describe the reactants. Instead, they tell you how many of each reactant molecule. Thus, the reactants are <u>O_2</u> and <u>H_2</u>.

b. (1 pt) Products appear on the right side of the arrow. The product is <u>H_2O</u>.

c. (1 pt) There are <u>two</u> H_2 molecules in the reaction, because of the "2" to the left of H_2.

7. (2 pts – one for the neutral pH and one for what is alkaline and what is acidic) The pH scale measures the acidity or alkalinity of a solution. On this scale, <u>7 is neutral. A pH lower than 7 is acidic, and a pH higher than 7 is alkaline</u>. The lower the pH the more acidic, and the higher the pH the more alkaline.

8. (1 pt) <u>Amino acids</u> link together to make proteins.

9. (1 pt) DNA stores information as a <u>sequence of nucleotide bases</u>, much like all of the English language can be stored as a sequence of dots and dashes in Morse code.

10. (5 points – one for each answer) The four stages are <u>glycolysis</u>, <u>the formation of acetyl coenzyme A</u>, <u>the Krebs cycle</u>, and <u>the electron transport system</u>. Glycolysis results in a gain of 2 ATPs, the Krebs cycle 2, and the electron transport system 32. Thus, <u>the electron transport system</u> produces the most energy.

11. (1 pt) The only stage that does not require oxygen is <u>glycolysis</u>.

12. Guanine and cytosine can bond together, as can adenine and thymine. In RNA, however, uracil replaces thymine. Thus when DNA has an adenine, RNA will have a uracil. When DNA has a thymine, RNA will have an adenine. When DNA has a cytosine, RNA will have a guanine, and when DNA has a guanine, RNA will have a cytosine. This makes the mRNA sequence:

a. (2 pts – $\frac{1}{3}$ for each base) <u>cytosine</u>, <u>guanine</u>, <u>uracil</u>, <u>uracil</u>, <u>adenine</u>, <u>cytosine</u>

b. (1 pt) It takes three nucleotide bases to code for an amino acid. Since this has six, it will code for two amino acids.

13. (4 pts – one for each stage) The stages in order are prophase, metaphase, anaphase, and telophase.

14. (1 pt) Diploid number is the total number of chromosomes in the cell. Haploid number is the number of homologous pairs. If there are a total of 16 chromosomes, then there must be 8 pairs. The haploid number is 8.

15. (1 pt) A gamete is haploid while a normal cell is diploid. This means that a gamete has only one chromosome from each homologous pair. A normal cell always both members of each homologous pair.

16. (4 pts – ½ for each stage) The stages are prophase I, metaphase I, anaphase I, telophase I, prophase II, metaphase II, anaphase II, telophase II.

17. (2 pts – one for each answer) Male gametes are called sperm. Female gametes are called eggs.

18. (1 pt) A virus is not alive, because a virus cannot reproduce on its own.

19. (1 pt) A vaccine is only good if you take it before getting infected, because it is meant to build up the antibodies that you need to fight the virus off before it overwhelms your body.

20. (3 pts – one for each letter. Specifying homozygous or heterozygous is worth ½, and the phenotype is worth 1/2)

a. This homozygous genotype results in a phenotype that is tall.
b. This heterozygous genotype results in a phenotype that is tall.
c. This homozygous genotype results in a phenotype that is short.

21. (3 pts – one for the parent genotypes, one for filling in the Punnett square correctly, and one for the proper percentages) One parent is homozygous dominant, so its genotype is "AA." The other is heterozygous, so its genotype is "Aa." The Punnett square looks like:

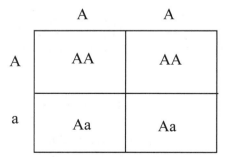

Thus, 50% of the offspring have the "AA" genotype and 50% have the "Aa" genotype. Since each offspring has at least one of the dominant allele, however, 100% have the axial flower phenotype.

22. (3 pts – one for the possible gametes, one for filling in the Punnett square correctly, and one for the proper percentages) Since the parents are both heterozygous in each allele, their genotypes are "SsYy." There are 4 possible gametes: *SY, Sy, sY, sy*. The resulting Punnett square, then, is:

	SY	*Sy*	*sY*	sy
SY	*SSYY*	*SSYy*	*SsYY*	*SsYy*
Sy	*SSYy*	*SSyy*	*SsYy*	*Ssyy*
sY	*SsYY*	*SsYy*	*ssYY*	*ssYy*
sy	*SsYy*	*Ssyy*	*ssYy*	*ssyy*

<u>smooth, yellow peas</u> (genotypes SSYY, SsYy, SSYy, SsYY) 9 of 16 or <u>56.25 %</u>
<u>smooth, green peas</u> (genotypes SSyy, Ssyy) 3 of 16 or <u>18.75 %</u>
<u>wrinkled, yellow peas</u> (genotypes ssYY, ssYy) 3 of 16 or <u>18.75 %</u>
<u>wrinkled, green peas</u> (genotype ssyy) <u>1 of 16 or 6.25 %</u>

23. (3 pts – one for the parent genotypes, one for filling in the Punnett square correctly, and one for the proper percentages) If the female is heterozygous, then her genotype is X^RX^r. Since the male is white eyed, his genotype is X^rY. The resulting Punnett Square is:

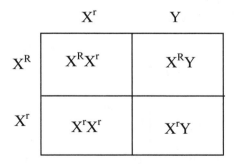

	X^r	Y
X^R	X^RX^r	X^RY
X^r	X^rX^r	X^rY

Thus, <u>50% of the females will be white-eyed and 50% of the males will be white-eyed.</u>

24. (1 pt) Not all traits are determined completely by genetics. Most are also determined by environmental factors and (in the case of humans) spiritual factors. <u>While the genetics are the same, the environmental and spiritual factors were probably different.</u>

25. (1 pt) The <u>X</u> chromosome should be studied. If it occurred on the autosomes, it should be roughly the same between men and women. If it occurred on the Y chromosome, it should also be roughly the same between men and women, since women have all of the alleles that are on the X chromosome. However, if it is on the X chromosome, men will have only one allele, making them more susceptible.

Total possible points: 51

SOLUTIONS TO QUARTERLY TEST #3

1. (1 pt) Macroevolution added <u>mutation</u> as a means by which information could be added to the genetic code.

2. (1 pt) <u>Darwin dispelled the idea of the immutability of the species.</u> By showing the evidence for microevolution, Darwin was able to show that species do adapt to a changing environment.

3. (1 pt) <u>He would say that since the transition from species to species takes such a short amount of time, there is virtually no chance of an intermediate link being fossilized.</u>

4. (2 pts – one for no, one for why) <u>This data provides strong evidence against macroevolution.</u> The similar structures are not a result of inheritance from a common ancestor, because <u>the similar structures are determined by quite different genes.</u>

5. (1 pt) <u>He did not recant.</u> Stories like that are not true.

6. (1 pt) Since the hypothesis of macroevolution postulates the giraffes are more directly related to horses than are fish or bacteria, macroevolution would predict that the cytochrome C sequence in a horse would be closest to that of a <u>giraffe</u>.

7. (1 pt) <u>It would not be as predicted.</u> The vast majority of protein sequences show no macroevolutionary relationships whatsoever.

8. (4 pts – one for the relationship, one for the description) Any two of the following will do. There could be more - check the book if you came up with one not listed here:

<u>The clownfish and the sea anemone form a mutualistic relationship. The clownfish is protected by the sea anemone, and it attracts food to the sea anemone.</u>

<u>The goby and the blind shrimp have a mutualistic relationship in which the goby protects the blind shrimp, and the blind shrimp provides a home for the goby.</u>

<u>The Oriental sweetlips and blue-streak wrasse form a mutualistic relationship in which the sweetlips gets it teeth cleaned by the wrasse and the wrasse gets food from the sweetlips' teeth.</u>

<u>Algae and fungi form a mutualistic relationship called a lichen, where the fungus provides shelter and protection for the alga and the alga provides food for the fungus.</u>

<u>Protozoa in the gut of a termite have a mutualistic relationship with the termite. The termite provides shelter and protection, while the protozoa digest cellulose.</u>

<u>Fungi and plant roots form a mutualistic relationship where the plant roots provide food for the fungi and the fungi provide minerals to the plant roots.</u>

9. (1 pt) <u>Oxygen is taken from the air principally by respiration.</u>

10. (1 pt) <u>Oxygen is restored to the air principally by photosynthesis.</u>

11. (1 pt) The smallest amount of energy is wasted where the biomass of one level is close to equal the biomass of the next level. Thus, <u>from primary consumer to secondary consumer wastes the least energy</u>.

12. (1 pt) <u>Global warming is not happening now</u>. The most reliable data indicate that any warming which did take place occurred before humans really started burning fossil fuels in earnest.

13. (1 pt) The vast majority of animals are <u>invertebrates</u>.

14. (1 pt) <u>Sponges get their prey by pumping water into themselves</u>. The water brings algae, bacteria, and organic matter that sponges eat.

15. (2 pts – ½ for each function) Amebocytes <u>help digest and transport nutrients</u>, <u>they help carry waste to be excreted</u>, <u>they bring necessary gases such as oxygen to the cells</u>, and <u>they form the spicules or spongin</u>.

16. (1 pt) <u>Cnidarians do not need these systems because their body walls are so thin that gases diffuse right through them</u>.

17. (1 pt) <u>It must be in medusa form</u>, because jellyfish can only reproduce sexually in medusa form.

18. (1 pt) <u>The earthworm will suffocate</u>, because oxygen cannot travel through a dry cuticle.

19. (1 pt) With complex nervous and digestive systems, it must need to seek out and fully digest prey. Thus, it is probably <u>free-living</u>.

20. (5 pts – one for each) <u>Exoskeleton</u>, <u>body segmentation</u>, <u>jointed appendages</u>, <u>open circulatory system</u>, and <u>a ventral nervous system</u>

21. (1 pt) <u>The injury gets sealed off to prevent bleeding, and then a new limb regenerates</u>.

22. (1 pt) They molt because <u>their exoskeletons get too small for their growing bodies</u>.

23. (5 pts – one for each) <u>Four pairs of walking legs</u>, <u>two segments in body</u>, <u>no antennae</u>, <u>book lungs</u>, and <u>four pairs of simple eyes</u>

24. (4 pts – one for each) <u>Three pairs of walking (or jumping) legs</u>, <u>wings</u>, <u>three segments in the body</u>, and <u>one pair of antennae</u>.

25. (1 pt) Insects do not need respiratory systems because of <u>a complex network of tracheas that allow air to travel throughout the body</u>.

Total possible points: 41

SOLUTIONS TO QUARTERLY TEST #4

1. (1 pt) The cerebrum integrates sensory information and coordinates the creature's response to that information.

2. (1 pt) Lampreys belong to class <u>Agnatha</u>.

3. (1 pt) The <u>red blood cells</u> carry oxygen in the blood.

4. (1 pt) Salmon and carp are both bony fishes. The <u>shark</u> is a cartilaginous fish and therefore has a more flexible skeleton.

5. (2 pts – one for each) This is <u>internal fertilization</u> because the fertilization takes place inside the female's body. It is <u>oviparous</u> development because the egg hatches outside the female's body.

6. (1 pt) Fish use their <u>lateral line</u> to sense vibrations in the water.

7. (1 pt) <u>Arteries carry blood away from the heart, while veins carry blood back to the heart</u>.

8. (1 pt) Meristematic tissue is made of undifferentiated cells. This means that the cells are relatively new, so <u>that portion of the plant is still growing</u>.

9. (5 pts – $\frac{1}{3}$ for each answer)

Letter	Shape	Margin	Venation
a.	Elliptical	Serrate	Pinnate
b.	Cordate	Entire	Pinnate
c.	Deltoid	Entire	Parallel
d.	Cleft	Dentate	Palmate
e.	Lobed	Entire	Pinnate

10. (1 pt) <u>The spongy mesophyll is typically on the underside of the leaf, and it is usually a lighter shade of green due to the fact that the photosynthesis cells are not as tightly packed there</u>.

11. (1 pt) A tree with an abscission layer is <u>deciduous</u>. Deciduous means the tree loses its leaves in winter. In order for the leaves to fall off, there must be an abscission layer that cuts the leaf off from the flow of nutrients.

12. (1 pt) <u>Phylum Bryophyta</u> contains the small plants, because bryophytes are not vascular. Thus, they must be small in order for nutrients to be able to reach the entire plant.

13. (1 pt) <u>The number of cotyledons produced in the seed</u> is the fundamental difference between monocots and dicots. Monocots produce one cotyledon, dicots produce two.

14. (1 pt) Water's <u>cohesion</u> is what forces it to travel upwards in a plant.

15. (2 pts – one for each) The structures are called <u>xylem</u> and <u>phloem</u>.

16. (1 pt) The <u>phloem</u> must be composed of living cells. Xylem are just tubes through which water flows, so the xylem cells need not be alive. Phloem cells actually aid in the transport of the organic material in the phloem, so those cells must be alive.

17. (2 pts – one for each) The <u>stamen</u> is the male reproductive organ, and the <u>carpel</u> is the female reproductive organ.

18. (1 pt) <u>A cotyledon provides nutrients to the embryo in the seed</u>.

19. (1 pt) <u>Insects aid in the reproductive process of flowering plants by transporting pollen from one flower to another</u>.

20. (5 pts – one for each) <u>covered with tough, dry scales</u>, <u>ectothermic</u>, <u>breathe with lungs throughout their lives</u>, <u>three-chambered heart with a ventricle that is partially divided</u>, <u>produce amniotic eggs covered with a leathery shell</u>

21. (1 pt) Order <u>Squamata</u> contains lizards.

22. (6 pts – one for each) <u>endothermic</u>, <u>heart with four chambers</u>, <u>toothless bill</u>, <u>oviparous, laying an amniotic egg which is covered in a lime-containing shell</u>, <u>covered with feathers</u>, and <u>skeleton composed of porous, lightweight bones</u>.

23. (1 pt) <u>When preening, a bird is actually oiling its feathers</u>. The feathers need to be oiled regularly to keep the hooked barbules sliding freely along the smooth barbules and to keep the feathers essentially waterproof.

24. (5 pts – one for each) <u>hair covering the skin</u>, <u>reproduce with internal fertilization and usually viviparous</u>, <u>nourish their young with milk secreted from specialized glands</u>, <u>four-chambered heart</u>, <u>endothermic</u>.

25. (1 pt) <u>Underhair</u> is used for insulation.

Total possible points: 45